# THE COMPETENCE PARADOX

# THE COMPETENCE PARADOX

*Leading People Through the Age of Algorithmic Management*

Ofem Ekapong Ofem

Published by Ideashaus Consulting, Minneapolis, Minnesota.

First Edition.

*For my family, and for humanity.*

# Contents

## PART FIVE: LEADING THROUGH PARADOX



## APPENDICES AND REFERENCE

# Author's Note

*The framework in this book rests on a systematic review of eighty-two peer-reviewed field studies and the paradox-theory scholarship that interprets them. The leaders and workers you will meet, including Marcus, Grace, Priya, and Carmen, are composites built from that evidence base rather than portraits of single individuals. The patterns they live through, the thresholds at which those patterns turn, and the levers that answer them are drawn from the research, not invented. Sources and effect sizes appear in the notes.*

# Preface

This is a book about a problem that hides in plain sight, inside the most data-rich organizations that have ever existed.

It is for any leader whose people are managed, even partly, by software. If a system in your organization sets goals, tracks work, scores performance, schedules shifts, routes tasks, or makes a recommendation that a human used to make, then you are leading inside the conditions this book describes. That now includes almost every leader, in almost every industry, which is why three executives from three unrelated fields could call me in a single quarter with what sounded like three different problems and turn out to be describing one.

The central claim is simple to state and uncomfortable to absorb. The systems that manage modern work build one kind of competence while eroding another, through the very same mechanism, at the very same time. They make people faster and more consistent while quietly thinning the judgment, the voice, the learning, and the trust that an organization depends on when the system is wrong. Worst of all, the erosion is invisible on the very dashboards that produce it, so the better your numbers look, the less you can see the cost. I call this the competence paradox, and once you learn to see it, you will find it everywhere.

The book is built in five parts. Part One shows you the problem and the evidence behind it. Part Two takes apart the four tensions that produce the paradox, one at a time. Part Three builds the five levers that answer them, which turn out to form a single self-reinforcing loop. Part Four puts the framework to work, from your first ninety days to the playbooks for your industry, the honest way to measure the second ledger, what to do when it goes wrong, and how to win the argument in the boardroom. Part Five is about the kind of leadership all of this asks of you, which is more demanding, and more human, than the kind the machine is replacing. The appendices give you the propositions to test in your own organization, the checklists to implement each lever, and a diagnostic to find which tension is hurting you most.

A word about the people in these pages. They are composites, drawn from eighty-two studies of real workplaces, and I have said so plainly wherever they appear. The stories are illustrations. The patterns, the thresholds, and the levers are not. Where I am standing on solid evidence I have tried to say so, and where I am extrapolating beyond it I have tried to say that too, because a book that asks you to act on its claims owes you an honest account of how far those claims can be trusted.

You can read this straight through, and I hope you will. But if you are in the middle of a fire, start with the tension that matches your symptom. If your best people are leaving, begin with specificity and autonomy. If your floor has gone quiet, begin with cadence and safety. If your bench is thinning, begin with efficiency and learning. If your people understand a system and still distrust it, begin with transparency and trust. Then turn to the lever that answers it, and start there.

The machine has taken over the easy half of leadership, the assigning and monitoring and scoring that it does faster and better than any human. What it has left behind is the hard half, the judgment and the voice and the developmental care that it cannot do at all, and that hard half was always the part that mattered. This book is about how to do it well. It is, in the end, an optimistic book, because the same forces that created the paradox have also made the human leader more necessary than ever. The only question is whether you will rise to the version of the job that remains.

PROLOGUE

# Three Calls, One Quarter

In the space of a single quarter, three people called me with the same problem, and not one of them knew it was the same problem.

The first was Marcus Bell, who ran operations for a logistics company with warehouses in three countries. His productivity numbers had never been higher. His turnover had never been worse. The analyst he had hired to explain the gap came back with a finding Marcus could not make sense of. The workers the system coached the most were the ones who quit the fastest. "We are identifying our strongest people," he told me, "engaging them harder than anyone, and then watching them walk out the door. Why would a system do that?"

The second was Dr. Grace Adeyemi, chief nursing officer of a twelve-hospital network. Her scheduling and patient-flow algorithms had cut average wait times by almost a fifth. She should have been celebrating. Instead she described something her nurse managers had started calling protocol rigidity. "Two years ago, if a nurse felt that something was off, she would walk down the hall and check," Grace said. "Now she checks the screen. I cannot tell anymore whether my people are getting safer or just getting quieter."

The third was Priya Raman, vice president of engineering at a software company that had rolled out AI coding tools eighteen months earlier. Her teams were shipping faster than they ever had. Her codebase was quietly coming apart. "We are putting up a taller building on a thinner foundation," she said. "The work looks better and it keeps getting more fragile. My senior engineers can see it happening. My junior engineers cannot even see what they have stopped being able to do."

Three industries. Three systems. Three problems that looked, on the surface, completely unrelated. A turnover problem, a safety problem, a quality problem. I had spent years reading everything I could find on what algorithmic systems do to the people who work beneath them, and I recognized the shape of all three before any of the callers had finished talking. They were not three problems. They were one problem wearing three uniforms.

• • •

Each of them had reached the same conclusion before they ever called me, and each had reached it wrongly. The system was broken, they assumed, and the fix was technical. Marcus needed better metrics. Grace needed a better interface. Priya needed a better model. So each had been pouring effort into tuning the machine, and each was quietly baffled that tuning it harder only seemed to sharpen the problem.

Here is what took all three of them months to see, and what this book exists to let you see on the first day instead of the hundredth. Their systems were not broken. The systems were working. They were doing precisely what they had been built to do, faster and more consistently than any human manager ever could. The damage was not a malfunction sitting next to the performance. The damage was the performance, viewed from its other side. The same feature that lifted Marcus's numbers was bleeding out his people. The same precision that cut Grace's wait times was teaching her nurses to stop looking up from the screen. The same speed that let Priya's teams ship was dissolving the judgment that kept the work sound.

I call this the competence paradox. It is the way an algorithmic system builds one kind of competence while it erodes another, through the very same mechanism, at the very same moment. It cannot be tuned away, because the gain and the loss are not two settings on a dial. They are one process seen from two angles. You do not solve a paradox. You learn to lead through it. And the leaders who learn how will quietly outperform the ones who are still trying to fix the machine.

What follows is the story of what Marcus, Grace, and Priya did once they stopped fixing and started leading, and of the framework the four of us built while they did it. By the last page you will be able to name the four tensions that produce the paradox, spot them in your own numbers before they ever reach your turnover report, and run the five levers that let people keep growing inside systems built to optimize them. We begin where Marcus began. On a Tuesday, with a number that was red.

PART ONE

# The Problem

CHAPTER ONE

# The Dashboard Boss

*A measure that cannot see the work will, given enough time, quietly destroy it.*

Marcus Bell found the problem on a Tuesday, and it was red.

He ran operations for fourteen warehouses and a little over two thousand people, and on the tablet in his hand every one of them had a color. Green was good. Yellow meant watch. Red meant act. The dashboard had cost a fortune and earned it back fast. Cycle times were down fourteen percent across the network. Scheduling errors had all but disappeared. The whole messy business of judging people had been reduced to a single number that refreshed every five minutes, the same number on the worker's screen and on his. Marcus had championed the system. He had stood in front of the board and defended the budget for it. He clicked between the color-coded profiles the way other men flip through photos of their children.

Then he landed on Carmen.

Eleven years on the floor. She had trained roughly half the people in the building. Her composite score was red.

Marcus asked her supervisor about it quietly, off to the side, so it would not become a thing. The answer was simple. Carmen's throughput had dipped during the previous quarter because she had spent three weeks onboarding new hires and untangling a conveyor reconfiguration that had jammed two shifts. The dashboard had no column for training three people. It did not weigh the conveyor she had kept running, or the institutional memory that lived only in her head, or the quiet way she settled a floor when a shift went sideways. It saw the hours she had poured into other people and scored them as the one thing she had failed to do, which was pick fast enough. The system was not wrong by its own logic. It simply could not see the work she was actually doing.

Here is the part that matters. Marcus was a good operator. He had come up off a floor himself, years earlier, and he knew exactly what a quarter like Carmen's looked like from the inside. Hand him her file ten years ago, the same dip in the same quarter, and he would have read the story in thirty seconds and walked over to thank her. But he did not have the file. He had the score. And the score recommended a performance improvement plan. So Marcus, trusting the expensive instrument he had fought to install, signed it.

Carmen resigned within eight weeks. She made no scene and filed no grievance. She found another position and left quietly, which is how the most valuable people usually leave. Six months later, new-hire retention in her unit had fallen by twenty-two percent. The three people she had been onboarding struggled without her, and one of them was gone within ninety days. The dashboard never connected her departure to the decline, because the metrics it tracked were too narrow to see what she had carried and too fast to notice what her absence cost. By the time the damage surfaced in a number Marcus actually watched, it was a year old and wearing a different name.

Marcus told me this story months later, in the flat voice of a man who has already done the arithmetic and does not enjoy the total. He was not careless. He was loyal. The trouble was that his loyalty had quietly transferred from his people to his instrument, and he had not noticed the moment it happened.

• • •

## When the System Becomes the Manager

This is not a story about a broken algorithm. The system worked exactly as designed. It measured what it was built to measure, and it measured it faster and more consistently than any human ever could. That is the trap, not the flaw. The problem was not a miscalibrated weight or a missing field that a better vendor could supply next quarter. The problem was that the system had quietly become the manager. And the manager it became could see throughput but not judgment, time on task but not the mentoring that made new hires productive, error rates falling in the work it tracked but not the capability rotting in the work it did not.

It is worth being precise about what "became the manager" means, because it is easy to wave the phrase around and miss the substance. When the new system arrived, it did not sit beside Marcus's supervisors as an extra opinion. It generated the roster. It triggered the performance review. It assigned the training module. It built the promotion shortlist and flagged the names for discipline. Each of those had once required a human being to observe a situation, interpret it, and decide. Now the deciding happened in software, continuously, at a scale no person could match, and the human supervisor was left to interpret the output. The system set the agenda. The supervisor responded to it. When Marcus went looking, he could not remember the last time one of his managers had overridden the dashboard rather than translated it. They had not stopped managing. They had started managing the screen.

Workplace monitoring is not new, and it would be dishonest to pretend otherwise. Frederick Taylor put a stopwatch on the factory floor more than a century ago. Call centers were recording and scoring conversations in the 1990s. Fleet tracking by satellite was ordinary in logistics by the middle of the 2000s. But the systems that spread over the past decade are different in kind, not just degree. The older tools observed. The newer ones decide. A scheduling algorithm does not lay out three options for a supervisor to weigh. It publishes the roster. A performance dashboard does not raise a concern for a manager to look into. It launches the intervention, and sometimes the manager learns about the concern at the same moment the worker does, or later. That shift, from a tool that informs a human decision to a system that makes the decision and informs the human, is the whole story compressed into a sentence.

The gap between what these systems can measure and what human capability actually consists of is the subject of this book. Carmen's red score is the entire argument in miniature. The dashboard made her unit faster and made it worse. It built one kind of competence, the consistent execution of a defined task, and it destroyed another, the seasoned judgment that holds a floor together, with the same hand and in the same motion. Marcus had called his version a turnover problem. In the same quarter, Grace was calling her version a safety problem and Priya was calling hers a quality problem. They were describing the same thing. It is not a bug you patch. It is a paradox you lead through.

## The Manager Becomes a Reader of Screens

There is a second casualty in this story, and it is easy to miss, because he is the one telling it. The dashboard did not only change Carmen. It changed Marcus.

When a manager spends his days interpreting a score instead of forming his own assessment, his judgment begins to follow the same arc as his workers'. Researchers who studied performance reviews found that when supervisors were handed an algorithm's suggested rating, their own ratings clustered into far narrower bands than when they judged without the suggestion. The system never forced a score. It merely offered one, and the simple presence of the number pulled human judgment toward it. Across more than three thousand reviews, the range of human assessment shrank. The managers believed they were exercising independent judgment. They were exercising the algorithm's judgment with a human signature on it.

Marcus recognized himself in that finding. He had not overruled the dashboard in months, and when he tried to recall the last time one of his supervisors had walked a red score back to green on the strength of something they had seen with their own eyes, he could not. His managers had not grown lazy. They had grown deferential to an instrument that was always confident, always fast, and never tired. The competence paradox does not stop at the front line. It climbs. The same system that made the worker better at the task and worse at the judgment did precisely the same thing to the people who were supposed to be judging the worker.

This climbing quality is what makes the paradox so hard to catch from the top. A senior leader looks down through layer after layer of people who are all, in good faith, reporting what the system tells them, and concludes that the system must be describing reality, because everyone is describing the system. The agreement is not evidence that the picture is true. It is evidence that everyone is looking at the same screen. Marcus had built a hall of mirrors and mistaken it for a window.

## The Rise of the Dashboard Boss

Algorithmic management is the use of automated systems to assign, monitor, evaluate, and sometimes discipline work in real time. The term covers a wide range, from a scheduling engine that rosters shifts against forecasted demand to a performance dashboard that scores employees against behavioral benchmarks updated every few minutes. What unites them is a single structural move. Decisions that once required a manager to watch, interpret, and act are now encoded in software that runs continuously, at scale, without the contextual understanding that an experienced manager brings to an ambiguous situation.

The scale of adoption is the part most leaders underestimate. Software dashboards now guide daily work in more than thirty countries, scoring and rewarding tasks as they happen. By 2022, more than forty percent of large companies were using predictive analytics in decisions about people, up from roughly twelve percent seven years earlier. Internal mobility engines, algorithmic performance ratings, and AI-assisted hiring tools are no longer experiments that early adopters pilot in a corner. They are infrastructure, wired into the operating systems of large organizations. The question for any leader is no longer whether these systems will arrive. They have arrived. The question is what happens to the people who work inside them, and how much an organization loses before it thinks to ask.

Marcus's network was one instance of a pattern that was running, that very quarter, in rooms he would never see. In Grace's hospitals, nurses were learning their schedules from an algorithm rather than from a charge nurse, and a triage system was ranking patients faster than any human could. In Priya's engineering organization, a tool was not only scoring the work but writing portions of it, suggesting the approach and refactoring code that no one had asked it to touch. Three different uniforms on the same body. Once you learn to see the shape, you find it everywhere, which is exactly why it is dangerous. A pattern that is everywhere is easy to mistake for the weather.

## Why It Spread So Fast

It is fair to ask why organizations adopted these systems so quickly and so completely, and the answer is not foolishness. The answer is that the systems solved real problems.

For a network like Marcus's, the old way of judging people was slow, inconsistent, and quietly unfair. One supervisor graded hard, another graded soft, and a worker's rating depended as much on which manager they happened to draw as on the work they actually did. The dashboard fixed that. It applied the same standard everywhere, at every hour, without the moods and blind spots that human managers carry. In an organization with thousands of people across a dozen sites, that consistency is not a small thing. It is worth real money and real fairness. Then a global disruption arrived, the work scattered, skilled people grew scarce, and the pressure to do more with fewer hands turned a useful option into a survival strategy. Timelines that might have taken years collapsed into months. The systems did not sneak in. They were invited, for good reasons, and the reasons were sound enough that almost no one stopped to ask what the invitation would cost.

## Three Rooms, One Pattern

Walk through the three rooms and the single pattern becomes hard to unsee. Start in logistics, where the system is most visible and most studied. Workers in a large distribution center carry scanners or wear devices that direct each movement and time it to the second. Productivity is rewarded automatically. Deviations are flagged without human review. The efficiency gains are real and well documented. Cycle times fall. Assignment becomes consistent. And yet the research on these environments keeps surfacing something the efficiency numbers never show. Workers report feeling constrained even when they are hitting every target. One study of drivers found that the system steered them to high-demand zones with impressive accuracy while stripping away the route discretion that experienced drivers had used for years to manage fatigue, protect their earnings, and build relationships with regular customers. Perceived freedom of action dropped measurably after as few as five prompts on a single task. The workers were performing better by the system's definition. By their own account, they were thinking less.

Move to Grace's hospitals. Predictive rostering assigns shifts against forecasted patient loads, fatigue indicators, and regulatory limits. The administrative burden on nurse managers drops, which frees real time for patient care. These are genuine gains, and Grace would be the first to defend them. But field research keeps finding the same second effect. When clinical teams are shut out of the design of these dashboards, the tools erode trust rather than build it. Show nurses exactly how the algorithm works, walk them through the weights, prove its accuracy on historical data, and their comprehension rises. Their trust does not. What moves trust is participation. When nurses help shape what the system measures and how it reports, trust climbs alongside understanding. When they are merely informed about a system designed without them, understanding breeds skepticism. They knew how the algorithm worked. They simply did not believe it understood what mattered in their corner of the floor. That is the protocol rigidity Grace described. The nurse stops walking down the hall to check, because the screen has already spoken, and the willingness to override a confident machine is the first thing to atrophy.

Now Priya's engineers. Productivity tools track time in applications and meetings and surface daily efficiency scores. Generative coding assistants go further. They do not only monitor the work. They complete portions of it, propose approaches, and in some configurations rewrite existing code without being asked. The gains are substantial and easy to celebrate. One study of AI-assisted software development recorded an eighteen percent rise in measurable productivity. Alongside it sat a decline in the deeper practices that build engineering judgment, the refactoring, the architectural review, the slow exploratory debugging that teaches a person how a system actually fails. The developers shipped more code in less time and learned less while doing it. The tool handled the routine work efficiently, which is its entire value, and in handling it removed the repetition that builds pattern recognition and the debugging intuition that senior engineers lean on when something breaks in a way no one predicted. The efficiency gain and the learning loss were not two outcomes to be managed separately. They were produced by the same tool, doing exactly what it was designed to do.

Three rooms. One pattern. In every case the same feature that produced the measurable gain also produced the developmental cost, and the two were welded together by a single mechanism. That is the thread that runs from Marcus's warehouse to Grace's ward to Priya's repository, and it is the thread we will follow for the rest of the book.

If there is a single principle to carry out of this chapter, it is this. What a system measures, it strengthens. What it cannot measure, it slowly starves. A leader's job, in the age of the dashboard, is to know the difference and to feed by hand what the machine cannot see.

## Why Tuning Makes It Worse

There is a useful distinction between a tool and an infrastructure, and the competence paradox lives in the moment the first becomes the second. A tool is something you pick up and put down. You choose when to use it, and you keep the judgment to override it when the situation calls for something the tool did not anticipate. An infrastructure is the ground you build on. It shapes what is possible before any single decision is made. When a system generates the roster, triggers the review, assigns the training, builds the shortlist, and flags the discipline, it has stopped being a tool that supports management. It has become the management layer itself.

This is why the natural first response fails. Confronted with the paradox, almost every leader reaches for the algorithm. Add variables. Adjust the weights. Refine the model. These moves are sensible and they help at the margins, which is exactly what makes them seductive. They do not touch the structure. A more sophisticated algorithm is still an algorithm. It still sets the terms, still decides what counts, still runs at a speed and scale that compress the space for human judgment. Slow the refresh from five minutes to fifteen and the worker's discretion is a little less compressed, but the underlying dynamic is unchanged. The system still defines the goals, still delivers the feedback, still applies the standard. The dial is not the problem. The structure is the problem.

The interventions that actually work, the ones we will build in Part Three, do not operate on the algorithm at all. They operate on the organization around it. They create deliberate space for human discretion inside an algorithmic environment, rather than trying to encode discretion into the algorithm and hoping the math will hold it. That is the single most important reframe in this book, and it is worth saying plainly before we go any further. You will not fix this with a better dashboard. You will lead through it with a better practice.

Marcus put it more bluntly than I would have. "I spent two years and a fortune teaching my best instrument to see everything," he said, "and it still cannot see the one woman who taught half my floor how to do the job." That is the sentence to keep. The most advanced measurement system in the building was blind to the most valuable work in the building, and no amount of additional precision was going to cure the blindness, because the blindness was the precision.

## Two Ledgers

Every algorithmic system keeps a ledger, and the trouble is that it keeps only one. On the visible ledger, Marcus's system was a triumph. Cycle times down fourteen percent. Scheduling errors gone. Granular, current, fair-seeming data on two thousand people, available on a tablet in real time. Every entry on that ledger was true.

There was a second ledger, and the system did not keep it, because it could not see the entries. The weeks Carmen spent making three new hires productive. The conveyor crisis she absorbed before it reached a customer. The retention that held in her unit because she was the reason people stayed, and the retention that collapsed once she left. The knowledge that walked out the door with her, and the ninety days it took for one of her trainees to follow her out. None of it appeared anywhere Marcus looked, because the instrument that had become his eyes was built to total the first ledger and had no column for the second. He was running his operation on half the books and calling it precision.

This is the quiet arithmetic of the competence paradox. The gains are measured, immediate, and celebrated. The costs are unmeasured, delayed, and absorbed somewhere the system never reports. By the time they surface in a number a leader actually watches, they have compounded for a year and lost their return address. An organization can post its best quarter and begin its worst decline in the same ninety days, and feel, the entire time, as though everything is finally under control.

## What the Dashboard Could Not See

The research that informs this book identifies four specific tensions that appear whenever algorithmic systems manage human work. Each one follows the same paradoxical shape. The system feature that produces a measurable gain also produces a developmental cost, and the two are joined by the same causal mechanism. It is worth naming them now, lightly, because they are the map for everything that follows.

The first is the tension between specificity and autonomy. Precise targets sharpen focus and lift accuracy, and the same precision compresses the discretionary space where professional judgment grows. This is Marcus's tension, and Carmen is its casualty. The second is the tension between cadence and psychological safety. Rapid feedback validates progress and confirms mastery, and the same constant visibility turns the system into a place where people manage how they look rather than how they work. This is Grace's tension, the one that taught her nurses to go quiet. The third is the tension between efficiency and learning depth. Optimization lifts throughput and crowds out the slow, exploratory practice that builds deep skill. This is Priya's tension, the thinner foundation under the taller building. The fourth is the tension between transparency and trust. Full explanation satisfies a person's need to understand and fails to satisfy their need to be recognized and heard, so understanding arrives without trust. This one runs through all three rooms.

Notice what happens when you line the four up. Each begins as a gift. A clearer target. Faster feedback. A more efficient path. A more open score. Nobody installs a system intending to compress judgment, silence concern, starve learning, or strand people in informed powerlessness. The harms are not the opposite of the gifts. They are the shadows the gifts cast, and they fall in exactly the places the system was never built to look. That is why careful leaders walk straight into this. They are not negligent. They are accepting the gift and failing to see the shadow, because the instrument that hands over the gift is the same instrument they have asked to do their seeing.

Stand back and the three phone calls resolve into a single picture. Marcus's turnover, Grace's safety scare, and Priya's decaying codebase were four tensions surfacing in three places, produced by the same structural feature of the systems all three had installed. None of them needed a better machine. They needed to see again what their machines had taught them to stop seeing.

That is the work of this book. Part One establishes the diagnosis and the evidence. Part Two takes the four tensions one at a time and shows the mechanism inside each. Part Three builds the five levers that answer them and shows how they fit together into a practice an organization sustains rather than a project it completes. Part Four puts that practice to work, from your first ninety days to the conversation you will have with your board. Part Five is about the kind of leadership the whole framework finally asks of you. Marcus did not need a smarter dashboard. He needed to walk back across the floor and see Carmen, and by the time he understood that, she was already gone. To understand why this happens to careful, competent leaders, and why it is structural rather than a personal failing, we have to leave the warehouse for a strange little driving experiment conducted in an English city more than fifteen years ago.

**THE EVIDENCE**

Carmen's case is not an outlier. A systematic review of eighty-two field studies found this dual effect, measurable performance rising while judgment, voice, or learning eroded at the same time, in more than two thirds of the corpus, across logistics, software, healthcare, finance, and retail.

It also carries thresholds. Perceived autonomy falls by roughly 0.40 standard deviations once dashboards refresh about every five minutes. People stop voicing concerns past roughly thirty feedback prompts an hour. Exploratory work drops about fifteen percent past twenty optimization nudges in a shift. Full methods, effect sizes, and sources appear in Chapter 3 and in the notes.

**MONDAY MORNING**

Pull one strong performer your dashboard currently rates poorly. Before you read the score, ask their manager a single question. What is this person doing that the system cannot see? Write the answer down. You have just found the first edge of your own competence paradox, and the rest of this book is about what to do with it.

CHAPTER TWO

# The Competence Paradox

*You cannot tune away a paradox. You can only learn to hold both ends of it at once.*

In 2008, a research team in Nottingham asked a question that sounds almost too simple to study. Does turn-by-turn navigation help people learn their way around a city? They took two groups of drivers and gave both the same set of routes through an unfamiliar area. One group used satellite navigation, the voice telling them where to turn. The other group used paper maps. After several trips, the researchers tested what each group actually knew.

The navigation group had arrived more efficiently. Fewer wrong turns, shorter travel times, less frustration along the way. They had also learned almost nothing about the city. When the researchers asked them to sketch the routes they had driven, to estimate the distance between two landmarks, or to find a new path when the usual road was closed, the navigation users performed worse than the map users on every measure. The tool that made them better at arriving had made them worse at knowing where they were. They could follow a route flawlessly. They could no longer build one.

That finding has been reproduced, in one form or another, for two decades, and it reaches far beyond driving. Cognitive scientists call the pattern offloading. When an external tool reliably handles a mental task, the brain stops spending energy to encode that information, because duplicating work the environment already does is wasteful. This is not a failure of attention or willpower. It is an efficient design choice by a system that conserves effort wherever it safely can. The same effect shows up with calculators and arithmetic, with spell-check and spelling, with search engines and plain factual recall. In each case the tool improves the immediate task and quietly reduces the underlying capability the task used to exercise. The knowledge that is never encoded is the knowledge that is not there when the tool is gone, when the situation shifts, or when the work demands something the tool was never built to do.

## From the Car to the Floor

The Nottingham study is clean in a way that real organizations never are. In a workplace the tool is not a satellite receiver, and the knowledge at stake is not a mental map of a city. The tool is a dashboard that sets goals, monitors performance, assigns tasks, evaluates quality, and delivers feedback in real time. The knowledge at stake is professional judgment. It is the ability to diagnose an unfamiliar problem, to recognize when the standard procedure does not fit the case in front of you, to weigh competing priorities when no scoring rubric applies, and to guide a colleague through a challenge that the system was never designed to anticipate. These are the capabilities that separate a competent professional from a person who can follow instructions.

Put the two together and you arrive at a conclusion that should make any thoughtful leader uneasy. The organizations investing most heavily in systems designed to improve performance are, in a measurable number of cases, degrading the very human capabilities that performance depends on over the long run. The systems are not failing. They are succeeding at what they were built to do, and that success is producing a second-order effect the systems cannot detect, because the effect lives in exactly the judgment, autonomy, and exploratory learning the systems were built to make unnecessary. This is not a contest between good technology and bad implementation. It is a structural feature of how algorithmic management meets human development. The same mechanism that produces the gain produces the loss.

You can watch the same trade play out on any rideshare platform. The app sends drivers to the busy zones with real accuracy, and the drivers who follow it earn more per hour than the ones who do not. They also learn less of the city, build fewer relationships in the neighborhoods they serve, and grow more dependent on the system for decisions they once made on their own. The immediate result is efficiency. The longer result is dependency, and the dependency stays invisible until the day the app is wrong or unavailable, at which point the drivers who kept their own map outperform the ones who traded it away. They had swapped knowledge for compliance, and the swap looked like a bargain right up until the moment it did not.

## Why the Early Days Feel Like Progress

There is a reason the competence paradox is so hard to catch while it is happening, and it has to do with how the early days feel.

When an algorithmic system takes over the cognitive chores of a job, the first thing a worker notices is relief. The ambiguity is gone. The expectations are explicit. The mental fatigue of deciding what to do next, of weighing this against that, of carrying the whole picture in your head, is suddenly the machine's problem rather than yours. People report higher satisfaction in the opening weeks of an algorithmic rollout, and they are not wrong to. Their cognitive load really has dropped. The work really is easier. This is the honeymoon, and it is genuine.

But the same relief that feels like a gift in week one is the mechanism of erosion by month six. The cognitive chores the system lifted away were not noise. They were exercise. Deciding what to aim for, judging your own progress, choosing how to approach a problem. Each one is a small repetition that keeps a capability alive, and each one feels, in the moment it is removed, like a burden lifted rather than a muscle going unused. By the time the loss is visible, the worker can no longer feel what is missing, because the faculty that would notice is the very faculty that has weakened. You do not feel yourself forgetting how to navigate while the voice tells you where to turn. You feel it the first time the voice goes quiet.

## Paradox, Not Problem

When I first laid this out for Marcus, he did what most operators do. He tried to make it an either-or. "So the dashboard is wrong," he said. No, I told him. The dashboard is doing its job. "Then it is right, and Carmen was underperforming." No again. The trap is the word in the middle of both sentences. The trap is or. Grace reached for the same word from the other direction, asking whether she should slow the system down or keep it fast. Priya asked whether she should trust the AI or pull it out. Every one of them was trying to choose a side of something that has no sides to choose.

It helps to be exact about language here, because the wrong word leads to the wrong fix. A trade-off is a situation where you buy one benefit at the cost of another. Spend more on speed and you give up some quality. The choice is unpleasant but clean, and you can reason your way to it. A dilemma is similar, two options each carrying a real cost. A contradiction is a logical impossibility, a thing that cannot be both true and false at once. A paradox is none of these. In the definition that organizational scholars Wendy Smith and Marianne Lewis made standard, a paradox is a set of interdependent elements that persist together, each sensible on its own, yet contradictory when you hold them side by side. The load-bearing word is interdependent. The opposing forces do not merely coexist. They produce one another. You cannot have the one without generating the other, and the tension does not resolve itself over time. It persists, which means it has to be managed rather than solved.

This is why the competence paradox is not a trade-off in which an organization simply chooses performance over development. If it were, the cure would be obvious. Dial the system back. The paradox describes a structure in which the very features that produce the performance gain also produce the developmental cost, through the same mechanism, at the same moment. Precision sharpens focus and compresses discretion in a single act. Rapid feedback confirms mastery and amplifies surveillance in a single act. Optimization lifts throughput and crowds out exploration in a single act. Transparency builds understanding and exposes the worker to judgment in a single act. These are not four problems waiting for four fixes. They are four expressions of one structural dynamic. The interdependence is what defeats the standard responses. You cannot find the right setting on the dial, because the gain and the cost are not on opposite ends of one continuum. They are produced together, by the same features, and they last as long as the features run.

## Four Tensions, One Structure

The evidence converges on four tensions, and it is worth meeting each one through the person who feels it most sharply, because that is how you will recognize it in your own building.

The first tension is specificity against autonomy, and it is Marcus's. Adaptive dashboards that refresh sub-goals every few minutes sharpen focus with real precision. One study recorded a twelve percent rise in task accuracy after such a system went live. The same study recorded a thirty percent fall in exploratory moves, the small experiments through which people discover better methods. Perceived freedom of action declined after as few as five prompts on a task. The mechanism is direct. The more precisely a system defines what to do, the less reason a worker has to exercise judgment about what to do. The act of defining the target is the act of compressing the discretion. Marcus saw it as drivers who could no longer find an address when the routing failed, and as a warehouse lead whose mentoring was invisible to the only eye that counted.

The second tension is cadence against psychological safety, and it is Grace's. Rapid feedback serves a genuine purpose. It validates progress, confirms mastery, and reduces uncertainty about where a person stands, and early on it can even raise people's sense of fairness. But the same cadence that validates also intensifies the feeling of being watched. One study found that dissent dropped by twenty-five percent once feedback passed roughly thirty prompts an hour. People stopped raising concerns, not because the concerns had vanished, but because the speed and visibility of the system made speaking up feel risky. Researchers call it a spotlight effect. The same act, making performance continuously visible, both confirms competence and triggers self-censorship. Grace saw it as a ward full of skilled nurses who had quietly decided it was safer to check the screen than to trust their own eyes.

The third tension is efficiency against learning depth, and it is Priya's. Optimization systems lift output by removing slack, standardizing the process, and steering people down the fastest path. The eighteen percent productivity rise in AI-assisted software work came with fewer refactoring cycles and less of the exploratory debugging that builds diagnostic skill. The same act, eliminating slack, both raises throughput and removes the unglamorous repetition through which expertise is actually built. The loop reinforces itself. Quick wins build confidence in the system's way, which reduces the felt need to explore, which compresses the time available to learn, which makes the system's way feel even more necessary. Priya saw it as a team that shipped beautifully until the day a problem arrived that the tool had never seen, and discovered that only the engineers who had learned the hard way could solve it.

The fourth tension is transparency against trust, and it runs through all three rooms. The common prescription for an opaque system is to open it up. Show people how they are scored and they will trust it. The evidence supports that only halfway. One analysis found that explainable systems raised trust by roughly half a standard deviation, but only when explanation came with voice. When workers could annotate, flag, or add context to the system's output, acceptance rose by forty percent. When they received explanation without any ability to respond, understanding went up and trust did not. The mechanism is dignity. Explanation answers a question of the head, do I understand what is happening. Voice answers a question of standing, am I a participant in this or merely its subject. A system that offers the first without the second produces a very particular kind of frustration, an informed powerlessness, and it is corrosive precisely because the person now understands exactly how little their judgment counts.

Four tensions. One structure. Look at them together and the family resemblance is unmistakable. In every case a single system feature delivers a measurable gain and an unmeasured cost in the same motion. They are not separate ailments. They are one condition seen from four angles.

## The Honest Objection

It is fair to push back here, and the strongest objection deserves a straight answer. Is this just nostalgia, a longing for slower days dressed up as analysis? Are we romanticizing the struggle of doing things the hard way?

No, and the distinction matters. The argument is not that struggle is virtuous or that the old methods were better. Plenty of the work these systems removed was tedious, error-prone, and worth removing. The argument is narrower and more uncomfortable. Some of the repetition the systems eliminated was not waste. It was the mechanism by which a capability stayed alive, and the system could not tell the difference, because from the outside the maintenance and the waste look identical. The afternoon a junior engineer spends writing a function the tool could generate in seconds looks like lost productivity, and on the visible ledger it is. On the second ledger it is practice in decomposition, in error handling, in design reasoning, the slow accumulation that turns a junior into a senior. Remove the afternoon and you remove the practice. The objection is right that not all friction is valuable. It is wrong to assume the system can sort the valuable friction from the wasteful kind. That sorting is a human judgment, and it is one of the first the system quietly takes over.

## The Surge and the Erosion

Underneath the four tensions sits a single mechanism with two faces, and naming the two faces is the fastest way to make the paradox usable. Call them the competence surge and discretion erosion. They are not two processes that happen to occur together. They are the same features doing the same things, viewed from two sides.

The surge is well understood and not in dispute. Algorithmic systems deliver clear goals, rapid feedback, and consistent standards, which are precisely the conditions that decades of goal-setting research identify as drivers of performance. Give a person a specific target, show them how they are doing against it in real time, and hold everyone to the same standard, and measurable performance improves. Algorithmic management supplies all three with a speed and consistency no human manager can match. The surge is real, and any honest account has to begin by granting it.

Erosion runs through the very same features. When the system sets the goal, the worker no longer practices the work of setting goals, which is where strategic thinking is built. When the system delivers constant feedback, the worker no longer practices self-assessment, which is where metacognition is built. When the system applies a consistent standard, the worker no longer practices judging what good work looks like in a particular context, which is where professional expertise is built. Each subtraction is small, and most of them feel like relief rather than loss. The system is carrying a cognitive burden that used to be tiring. But each subtraction also removes a task that, when a person performs it, builds a specific capability. Defining your own goals, assessing your own work, judging your own quality. These are not optional extras layered on top of doing the job. They are the substrate of competence. When the system handles them, task performance can rise while the deeper capability quietly stalls. It is the Nottingham drivers again. They arrived faster and learned less about the territory.

Grace gave me the cleanest description of this I have heard. "My best nurses used to carry a question in their heads at all times," she said. "Does this patient match the pattern, or is this one different? The system answers that question for them now, and it is usually right. But a nurse who has not asked herself that question in a year has stopped being the kind of nurse who can answer it on the day the system is wrong." That is discretion erosion in a single breath. The skill did not fail. It went unused, and unused skills do not wait politely for the day you need them again.

The point worth holding is that the surge and the erosion come from the same act. Specificity sharpens focus and compresses discretion in the one act of defining the target. Feedback confirms mastery and amplifies surveillance in the one act of making performance visible. Optimization lifts output and starves exploration in the one act of removing slack. That is what makes this a paradox in the technical sense rather than a slogan. The interdependent elements produce each other. You cannot get the surge without generating the erosion, because they share an origin. And it leads to the conclusion that reorganizes everything. A system can be working and harming at the same time. The work is not to fix a broken machine. The work is to supplement a working one with the capabilities it cannot provide on its own.

## What Managing a Paradox Actually Looks Like

Managing a paradox sounds abstract until you have done it once. Here is the shape of it.

You do not pick precision or autonomy. You keep the precise targets, because they work, and you build a recurring forum where the people doing the work help decide which targets are right, so the precision serves their judgment instead of replacing it. You do not pick cadence or safety. You keep the fast feedback, and you carve out a protected pause where a human helps the worker interpret the signal rather than merely react to it. You are not splitting the difference. You are holding both ends at full strength and building the structure that lets them coexist. The tension never disappears. On a good week it is productive. On a bad week it tips, and you catch the tip early, because you are finally watching for it. That is dynamic equilibrium. It is less like solving an equation and more like keeping your balance on a moving deck. You do not finish. You stay upright, and you get better at it.

## Five Levers, Not One Switch

Paradox theory offers a specific prescription for tensions that cannot be resolved. Manage them. That sounds vague until you see what it means in practice. It does not mean finding a comfortable midpoint between the two poles. It means building structures that let both poles operate at once, watching the balance between them, and adjusting when one side begins to dominate. Scholars call the goal a dynamic equilibrium. It is not a static compromise. It is an ongoing practice of holding competing demands in productive tension.

For the competence paradox, that means preserving the genuine performance gains of algorithmic management while actively rebuilding the developmental space those same systems compress. The research supports five levers, and I will only name them here, because Part Three gives each one a chapter and a real implementation. The first is goal calibration, which brings workers into the setting of the metrics the system optimizes. The second is hybrid coaching, which inserts a protected human pause into the feedback loop so people interpret signals together rather than reacting to them alone. The third is rotational upskilling, which deliberately puts people back into work the system has automated away, so the underlying skill does not vanish. The fourth is explainable dashboards, which turn disclosed logic into understood logic the worker can weigh against experience. The fifth is moderated voice forums, which give professional judgment a structured channel to shape the system's ongoing design. The levers are sequenced on purpose. Each one builds the capacity the next one needs, and together they form a loop the organization sustains rather than a project it finishes.

## What We Do Not Yet Know

Honesty about the edges of the evidence is not a weakness in a framework. It is what makes the framework worth trusting. Three things are worth saying plainly before we go on.

First, the thresholds are patterns, not laws. Five-minute refresh rates, thirty prompts an hour, twenty nudges a shift. These are useful markers drawn from a large body of studies, not precise cutoffs that apply identically everywhere. Boundary conditions move them. An organization with high comfort around technology and a strong coaching culture may absorb a feedback intensity that would be corrosive somewhere else. A team with little appetite for learning may show erosion well below the published thresholds. Treat the numbers as places to start looking, not as settings to dial in.

Second, the costs are hard to measure, which is part of why they are so easy to ignore. The damage shows up in capabilities that resist quantification, in judgment and adaptability and the willingness to raise a concern no one asked for. The evidence relies on proxies, exploration frequency, dissent rates, self-reported autonomy, that capture pieces of these capabilities without measuring them directly. Better instruments would help. But the very difficulty of measuring these capabilities is part of why they are vulnerable in the first place. A system that optimizes what it can measure will always compress what it cannot.

Third, the time horizon is genuinely open. Most of the studies are short, weeks to months rather than years. We do not yet know whether discretion erosion accelerates as people grow more habituated to algorithmic direction, or stabilizes once a baseline of dependence sets in, or whether the levers hold their effect over the long run without periodic recalibration. These are not reasons to wait for perfect evidence, because perfect evidence is not coming. They are reasons to use what we have while building the organizational habit of generating our own.

Marcus heard the same thing in his own language. "So the fix is not a better dashboard," he said slowly, working it out at the table. "The fix is everything around the dashboard." That was the morning he stopped being a man with a turnover problem and became a man with a design problem, which is a far better problem to have, because design problems have answers.

Marcus, Grace, and Priya each left this conversation with the same uncomfortable relief. The problem was larger than they had thought, because it was structural and would not fix itself. It was also simpler than they had thought, because it was one problem rather than three, and it had a shape they could now name. What none of them would take on faith, and what no leader should, is a claim this large without an evidence base underneath it. So the next chapter opens that base, and answers a question that should bother every executive who reads it. If this pattern is real and this consistent, why does it stay hidden inside organizations that have more data than any in history?

### THE EVIDENCE

The four tensions were not chosen for elegance. In a review of eighty-two studies, a tension qualified only if it appeared in at least forty percent of the corpus and showed benefit and harm at the same time. All four cleared the bar. Specificity against autonomy appeared in fifty-eight percent of studies, cadence against safety in forty-seven, efficiency against learning in forty-four, transparency against trust in forty-two.

The two-faced mechanism has a research lineage worth knowing. The surge rests on goal-setting theory, the erosion on self-determination theory and the job demands-resources model, and the whole is held together by paradox theory. Citations appear in the notes.

### MONDAY MORNING

Take the four tensions to your next leadership meeting and assign each one an owner who will watch a single leading indicator for it. Autonomy, voice, exploration, trust. You are not solving anything yet. You are building the second set of eyes the system does not have, which is the precondition for everything in Part Three.

CHAPTER THREE

# The Evidence Base

*A single case persuades no one. A hundred, gathered honestly, leave nowhere to look away.*

Grace Adeyemi did not believe me, and she was right not to.

She had been a scientist before she was an executive, and she ran her twelve hospitals the way she had once run a lab, on evidence she could inspect. She let me finish the whole argument, the surge and the erosion, the four tensions, the mechanism, and then she set down her pen and said the thing I had been waiting for someone to say. "You have told me three stories. Marcus, Priya, and a nurse on one of my own floors. Three stories are not data. They are three stories. Show me you are describing the world, and not just the three people who happened to call you in the same quarter."

It was the correct objection, and I told her so. The plural of anecdote is not evidence, and a consultant with a theory can always find three cases that flatter it. She wanted the systematic review, the part of the work that sits behind the stories, and she was entitled to it. Because the honest answer to her challenge turns out to be the same as the answer to a question that should bother every executive who has read this far. If this pattern is real, and this consistent, why does it stay hidden inside organizations that collect more data about their people than any institutions in human history?

## Why You Cannot See It From Inside

The reason the paradox hides is not that the data is missing. It is that the data is pointed in the wrong direction.

Start with what the instruments are built to record. A dashboard counts throughput, cycle time, error rates, response times, the quantities that move when performance improves. There is no field anywhere on it for judgment that went unexercised, for the concern a nurse decided not to raise, for the mental map a driver never built. The surge is captured in high resolution because the surge is what the system was designed to measure. The erosion lives in the negative space the instrument was never built to see. You cannot find what you are not recording, and no organization records the slow disappearance of a capability that still, for now, appears to be present.

Then there is survivorship. The people the system damages most are often the ones who leave, and they leave the way Carmen left, quietly and without a scene. Their departures are logged as ordinary attrition, scattered across many months, and attributed to the usual suspects, pay or commute or a better offer. The signal is real, but it is dispersed across the calendar and mislabeled at the exit interview, so it never assembles itself into a pattern anyone can see. The organization keeps a careful record of who left and almost no record of why it mattered.

Add the lag. The gains arrive immediately and the costs compound slowly, and by the time a cost surfaces it has changed its name. It shows up as a quality incident, a safety event, a hiring crisis, a project that fails in a way no one can quite explain. Cause and effect are separated by enough months that no one thinks to connect the decaying capability to the impressive dashboard that decayed it. The two live in different quarters and different reports, and nobody owns the line between them.

Finally there is aggregation, which is the quiet killer. Averages hide the paradox almost perfectly. A unit's numbers look healthy because the measured surge offsets a developmental loss that has no column of its own, and the loss is invisible until you disaggregate and look at the right variables in the right subgroups. The dashboard is built to roll everything up into a clean composite, which is exactly the operation that makes the damage disappear. Grace recognized every one of these from her own hospitals before I had finished listing them. Her data was not absent. It was facing the wrong way.

## The Dashboards Were All Green

Grace gave me her own example before I could give her one, which is how I knew she had already half believed me.

In her network, she said, every operational dashboard had been green for eighteen months. Wait times down, bed turnover up, staffing costs controlled, patient satisfaction holding steady. By every measure the executive team reviewed, the algorithmic rollout had been a triumph, and the board had been told exactly that. And in the very same eighteen months, the thing she could not put on a slide had been spreading across her floors. Nurses checking the screen instead of the patient. Veterans deferring to a system that did not know the woman in room nine the way they did. A slow, quiet narrowing of the clinical judgment that had always been her network's real safety system. None of it had a dashboard, so none of it had reached the board, so officially it was not happening.

That is the trap stated as cleanly as I have ever heard it. The richer your data, the more confident you become that you can see, and the more invisible the things your data does not capture become, because the absence of a number gets read as the absence of a problem rather than the absence of a measurement. An organization swimming in data does not suffer from too little information. It suffers from too much confidence in the information it happens to have. Grace had twelve hospitals of evidence that everything was fine and one growing dread that it was not, and the evidence was louder than the dread, which is why she had called me instead of acting.

## How We Looked

So we did the unglamorous thing. We stopped collecting stories and went looking for everyone who had already measured this, in real workplaces, with real instruments, and we did it in a way designed to keep us honest.

A systematic review is not the same as citing the studies that happen to support your hunch, which anyone can do and most people do without noticing. The discipline is that you write down your search rules and your inclusion criteria first, before you see the results, and then you search exhaustively and screen every record against the rules you set in advance. You are not allowed to quietly drop the inconvenient studies, because you committed to the rules before you knew which studies would be inconvenient. It is a procedure built specifically to defeat the human talent for finding what we hoped to find.

We searched the major research databases for field studies of algorithmic management and its effects on the people working under it. The initial search returned thousands of records, and most of them fell away against the criteria, which were strict and worth stating, because the strictness is what separates this from a folder of clippings. A study had to take place in a real workplace, not a laboratory simulation with students role-playing under fluorescent lights. It had to involve an actual algorithmic management system, not merely the presence of software. And it had to measure at least one performance outcome and at least one human outcome in the same setting, autonomy or learning or voice or trust or wellbeing, so that both sides of the ledger were visible in the same place at the same time. A study that measured only productivity, or only sentiment, could not speak to a paradox that is about the relationship between the two, and was set aside.

It is as important to say what we threw out. We set aside the laboratory studies, however elegant, because a student clicking through a simulated task for course credit is not a nurse at three in the morning or a driver at a flooded bridge, and the stakes change the behavior. We set aside the opinion pieces and the vendor white papers, which have a reliable habit of finding only good news. We set aside the studies that measured a single side of the ledger, because a paper showing only a productivity gain cannot tell you what that gain cost, and a paper showing only sagging morale cannot tell you what was bought with it. Each exclusion shrank the pile and sharpened it. What remained was not everything written about these systems. It was everything written that could actually speak to the question of both sides at once.

What survived the screening was eighty-two studies, spanning logistics, software, healthcare, finance, and retail, published across the past decade. It is worth being precise about what that is and is not. It is a synthesis of other people's field measurements, not a single new experiment of our own, and that is both a limitation and a strength. No single organization, no single industry, no single research team, no single method. Eighty-two independent looks at the same animal from eighty-two different angles. When a finding repeats across that many settings, gathered by people who were not coordinating and often were not even asking the same question, you are no longer looking at a quirk of one workplace. You are looking at something about the way this kind of system meets this kind of worker.

## The Shape of a Single Study

It helps to see what one of these eighty-two studies actually looks like, because a number like sixty-eight percent can float free of the world it came from.

Take one of the logistics studies, the kind that anchors the warehouse end of the corpus. Researchers followed several hundred order-pickers in a large distribution center over a number of months. The workers wore devices that directed each pick and timed it, and the researchers had access to the performance data the system already generated, throughput, error rates, the familiar quantities. To that they added what the system did not collect. They surveyed the workers on their sense of autonomy and their willingness to suggest improvements, and they tracked who stayed and who left.

The performance side told the triumphant story everyone expected. Throughput up, errors down, new pickers productive faster than ever. The other side told the story no dashboard had been built to tell. Perceived autonomy had fallen, the rate of worker-suggested improvements had dropped almost to nothing, and the strongest performers, the experienced pickers the system rated highest, were leaving at a higher rate than the middling ones. One study, one warehouse, both ledgers measured in the same place at the same time. Now imagine that same result arriving from a hospital in one country, a bank in another, a software team in a third, eighty-two times, in settings that shared nothing except the shape of the management. The single study makes the pattern real. The eighty-two make it undeniable.

## What Eighty-Two Studies Agree On

Four findings came back so consistently that they ended up organizing the rest of this book.

The first is that the dual effect is the rule, not the exception. In more than two thirds of the studies, a measurable performance gain sat directly beside a measurable human cost in the same setting, produced by the same system. This was not a handful of cautionary tales surrounded by clean successes. In the clear majority of carefully measured cases, the gain and the cost arrived together. That alone should change how a leader reads a glowing dashboard, because it means the glow is not the whole picture roughly two times out of three.

The second is that it crosses industries. The warehouse, the hospital ward, the trading desk, the software team, the retail floor. The work could not be more different, and the shape of the effect is the same in all of it. This is why a logistics operator, a chief nursing officer, and a head of engineering could call me in one quarter with what sounded like three unrelated problems and turn out to be describing one. The pattern does not belong to an industry. It belongs to a way of managing.

The third is that the effect concentrates into four tensions, and they are not equally common. We counted a study toward a tension only when it showed both the benefit and the harm of that specific pairing. The tension between specificity and autonomy appeared in fifty-eight percent of the studies, the most common by a clear margin. Cadence against psychological safety appeared in forty-seven percent. Efficiency against learning depth in forty-four. Transparency against trust in forty-two. Those four are not a tidy framework imposed on messy data. They are where the data piled up, and they are the spine of Part Two.

The fourth finding is the most important in the book, and Grace sat up when I reached it. The harms are conditional, not fixed. The same kind of system that gutted autonomy in one workplace left it perfectly intact in another, and the difference came down to identifiable conditions. How fast the system ran. How ready the workforce was for the technology. How much human coaching surrounded it. How much voice people had in shaping it. Conditional means changeable. If the harm were an iron law of the technology, leadership could do nothing useful but unplug the machine and accept the lost performance. It is not an iron law. It is a function of design choices that leaders already control, which means there is real work for a leader to do, and the rest of this book is that work.

Put those four findings side by side and the shape of the whole problem appears. It is common, not rare. It is universal across industries, not peculiar to one. It organizes into four recognizable tensions rather than a hundred scattered complaints. And it is conditional, which means it answers to leadership. A problem that is common, universal, structured, and changeable is the best kind of problem an executive can ever be handed, because it is big enough to matter and shaped enough to act on. Grace heard all four findings and drew the obvious conclusion before

I could state it for her. This was not a reason to tear out her systems. It was a job description.

## Five Sectors, One Disease

The same shape wore five very different costumes, and it is worth seeing each one, because you will recognize your own industry in the list and perhaps be unsettled by how familiar the others sound.

In logistics, it looked like Marcus's warehouse and Daniel's closed bridge. Precise direction, soaring throughput, and the slow death of the route knowledge and floor judgment that used to live in the workers' heads. In software, it looked like Priya's codebase. Faster shipping, more output, and a generation of engineers who could accept what a tool produced but were losing the ability to evaluate it. In healthcare, it looked like Grace's wards. Shorter waits, smoother flow, and nurses who had begun to trust the screen over their own trained eyes.

In finance, it looked like analysts and traders leaning on algorithmic signals, sharper and faster in ordinary markets and strangely lost when the market did something the model had never seen, which is precisely when judgment is worth the most. In retail, it looked like scheduling and task systems that ran stores with fewer managers and less waste, while the floor staff who once knew their regulars and their stock by instinct became interchangeable executors of a screen's instructions. Five industries. Five sets of numbers that improved. Five capabilities that quietly thinned. One disease, five costumes.

Grace stopped me there. "You are telling me this is not a healthcare problem," she said. "It is not," I told her. "Your nurses are not the exception. They are the version of the rule that happens to wear scrubs." She did not like the answer. But she stopped arguing with the data, which is the most any evidence can ask for.

## The Theories Underneath

A pattern that repeats this reliably needs an explanation, or it is just a coincidence with excellent attendance. Four established theories, taken together, account for both the surge and the erosion, and you do not need any of the jargon to use them.

The first explains the surge. Decades of research on goal-setting establish that specific, challenging goals paired with regular feedback drive performance, reliably and across almost every kind of task. Algorithmic systems supply specific goals and relentless feedback as their core function. So the surge is not a surprise or an accident. It is exactly what the most robust finding in the field predicts will happen when you give people clear targets and constant information about their progress. The machine is a goal-setting engine, and goal-setting works.

The second explains the erosion. Research on human motivation finds that people need three things to stay engaged and to keep developing. A sense of autonomy, meaning some real say over how they work. A sense of competence, meaning the experience of getting better. And a sense of relatedness, meaning genuine connection to other people. An algorithmic system can feed the competence need handsomely, with its clear signals of progress, while it quietly starves the autonomy need, because it has taken over the how. That split is the felt experience of the erosion. People report feeling more capable and less free at the same time, and the slow draining of their own initiative is the autonomy need going hungry while the competence need is fed.

The third explains the thresholds. A long line of work on job design holds that every feature of a job is either a resource that energizes or a demand that depletes, and that the very same feature can be one or the other depending on the dose. Feedback in moderation is a resource. The same feedback at thirty prompts an hour becomes a demand that exhausts and intimidates. This is why the harms have thresholds rather than being present or absent. The dose makes the poison, and the dose is something a leader sets.

The fourth explains why none of it goes away on its own. Work on organizational paradox holds that certain tensions are permanent and interdependent, that the poles produce each other, and that such tensions are managed rather than solved. This is the frame that holds the other three together and tells a leader what kind of problem they are facing, which is the most useful thing a theory can do, because the wrong diagnosis guarantees the wrong treatment. Put the four in a single sentence and you have the engine of this book. The systems work because of goal-setting, they harm because they violate the need for autonomy, the harm scales with dose, and the whole thing persists because it is a genuine paradox.

Grace summarized the four theories back to me in a sentence I have stolen ever since. "So the machine is a brilliant coach and a terrible mentor," she said. "It will tell you exactly how you are doing, all day long, and it will never once ask you what you think." That is the motivation research in the language of someone who has actually run a floor. The coaching feeds competence. The silence starves autonomy. A person fed on coaching and starved of voice will perform beautifully and develop not at all.

## Why Smart Leaders Argue With the Data

There is one more reason the pattern hides, and it is the most human of them. Leaders do not want to see it.

By the time a leader is shown this evidence, the system is usually theirs. They championed it, defended its budget, staked their judgment on it, and watched it deliver the very gains they had promised. To accept that the same system is also doing quiet harm is to accept that the triumph is complicated, and complicated triumphs are hard to carry back to a board that wanted a clean win. So the evidence meets a wall of motivated reasoning, not because leaders are foolish but because they are invested, and the most invested are usually the most capable, which is why this catches good leaders more reliably than poor ones. Marcus did not miss Carmen because he was careless. He missed her because he had spent two years believing in the instrument that was hiding her.

This is why the evidence has to be falsifiable and why the predictions have to be stated in the open. A leader invested in a system will not be moved by a story, because they already have a better story, the one glowing on the dashboard. They can sometimes be moved by a specific, checkable prediction about their own operation, run on their own numbers, that turns out to be true. So the point of the next four chapters is not to win an argument. It is to hand you tests you can run yourself, because the only evidence that reliably changes an invested mind is the evidence it gathers with its own hands.

## What Would Prove Us Wrong

A claim that cannot be wrong is not worth much, and Grace would have been the first to say so. So before we go further, here is what would falsify this framework, stated as specific predictions you can test against your own operation.

From the four tensions we derived four falsifiable propositions, each naming a condition and a measurable consequence. The first concerns specificity and autonomy. Where dashboards refresh roughly every five minutes and the workforce is not ready for the technology, perceived autonomy should fall by about four tenths of a standard deviation, while the sense of personal competence rises by about three tenths, and where the workforce is technologically ready, the autonomy loss should largely disappear even at that fast cadence. The other three make similar bets, about feedback cadence and safety, about optimization and learning, about transparency and trust, and each gets its own chapter and its own test.

The point of writing them this way is discipline, not decoration. If you run these systems at these intensities under these conditions and you do not observe these effects, then the framework is wrong and you should put it down. We state the predictions in the open precisely so that they can be checked and, if they fail, discarded. That is the difference between a theory and a sales pitch. A sales pitch explains every outcome after the fact. A theory tells you in advance what it forbids. Part Two takes the four tensions one at a time, and for each it lays out the prediction, the mechanism that produces it, and the answer a leader can actually build.

Grace listened to all of it without writing anything down, which from her was a form of close attention. When I finished she was quiet for a moment. "All right," she said. "It is real. Now tell me what to do about the one that is killing my floor." We will get to hers, and it is a strange and dangerous one. But we start with the most common of the four, the tension that appeared in more studies than any other and the one that cost Marcus his best warehouse lead. The tension between telling people exactly what to do and leaving them the room to become good at it.

**THE EVIDENCE**

The review followed a structured scoping method: search rules and inclusion criteria fixed in advance, exhaustive search of the major databases, every record screened against the criteria. Of several thousand initial records, eighty-two field studies met the bar of a real workplace, a real algorithmic system, and at least one performance outcome and one human outcome measured together.

Headline result: the dual effect appeared in more than two thirds of studies, across five sectors, concentrating into four tensions at fifty-eight, forty-seven, forty-four, and forty-two percent. Crucially, the harms were conditional on cadence, readiness, coaching, and voice. Full method and sources are in the notes.

**MONDAY MORNING**

Pick one team running an algorithmic system. Find two numbers the system already collects, a performance number that is up and to the right, and anything at all that proxies a human outcome, voluntary attrition of strong performers, near misses, declined suggestions. Put the two on the same page for the first time. The paradox lives in the gap between them, and most organizations have simply never drawn the two lines on one chart.

PART TWO

# The Four Tensions

CHAPTER FOUR

# Precision and Autonomy

*Tell people exactly what to do, and you will get exactly that, and nothing you did not think to ask for.*

The bridge was out, and Daniel did not know the city.

Daniel drove for one of Marcus's regional carriers. Three years on the job, and every mile of those three years had come with the routing app talking in his ear, telling him which turn, which lane, which order to make his stops. By every number that mattered he was a good driver. On time, on route, low on fuel, high on the dashboard. On the morning the river flooded and the city closed the main bridge, the app had not yet caught the closure, and it kept routing him to it. He drove to the barricades, turned around, and the app sent him back to the same barricades, because following the app was what he had always done and what he was rated on. He sat at the closed bridge and called dispatch, and he waited, because he did not have another way to the other side of the river in his head.

Two trucks behind him was Rosa, who had been driving for twenty-two years and had come up before the app existed. She heard the bridge was out on the radio, rerouted herself through a maze of side streets she had known for two decades, made her deliveries, and was halfway through her afternoon by the time

Daniel got moving again. Same company, same tool, same closed road. One driver had a city in his head. The other had an app, and that morning the app was wrong.

What happened to Daniel afterward is the part Marcus could not stop thinking about. Daniel was not embarrassed. He did not believe he had failed at anything, because by his lights he had done exactly what he was supposed to do. He had followed the system, the system had been wrong, and so the lateness was the system's fault and not his. He was right, in a narrow way, and that narrow rightness was the whole problem. The system had trained him so thoroughly that he no longer experienced its failures as his responsibility, which meant he had no reason to build the backup judgment that would have let him recover. Why learn the city, when knowing the city is no longer your job?

Marcus told me this as a near miss rather than a disaster, because nothing actually broke. The deliveries were late and the customers were annoyed and that was the end of it. But it frightened him more than his turnover numbers had, because it showed him the shape of a future he was building on purpose. A fleet of Daniels, superb on the ordinary days when the system is right, helpless on the rare day when it is not, and no Rosas coming up behind them, because the same system that was so efficiently producing Daniels had quietly stopped producing Rosas. He was not training drivers anymore. He was training followers, and followers are wonderful until the thing they follow leads them to a barricade.

## Two Kinds of Knowing

There are two ways to know how to do a thing, and the difference between them is the whole of this chapter.

The first is knowing how to follow a correct instruction. The second is knowing why the instruction is correct, well enough to recognize when it has stopped being correct and to build a new one on the spot. The first kind is fast to acquire and fragile under surprise. The second kind is slow to build and holds up when the world departs from the script. Rosa had the second kind. Daniel had the first. On every ordinary day the two are indistinguishable, which is exactly why the difference is so easy to stop investing in. Why spend months growing the slow, durable kind of knowing when the system can install the fast, fragile kind in an afternoon, and the numbers look the same either way?

The answer is that the two kinds of knowing only look the same until the moment they do not, and that moment is never on the schedule. It is the flooded bridge, the patient who does not match the pattern, the bug the tool has never seen. Precision systems are extraordinary at manufacturing the first kind of knowing at scale and at speed. What they cannot do, and what they quietly discourage, is the building of the second kind, because the second kind grows only through exactly the struggle, the deciding and the getting-it-wrong, that the system exists to eliminate. You do not get the deep map by being told every turn. You get it by occasionally having to find your own way, which is the very thing a good precision system is designed to make unnecessary.

## The Surge Is Real

Begin, as always, by granting what works, because if you pretend the precision has no value you will lose the argument with the first operator who has watched his numbers climb.

Specific targets sharpen performance, and the effect is not small. In one field study, an adaptive system that refreshed precise sub-goals every few minutes lifted task accuracy by twelve percent. Onboarding collapses from months to days, because the system carries the knowledge a new hire used to have to accumulate the slow way. Errors fall. Variation between workers narrows. Marcus's network gained every bit of this, and the gains are what paid for the system in the first place. The routing app made Daniel a competent, productive driver in a fraction of the time it had taken Rosa to become one. That is not nothing. For a business that needs to put thousands of capable people on the road quickly, that is close to everything, and any framework that waves it away deserves to be ignored.

The same gift shows up in Priya's world, which is worth noting so that no one mistakes this for a story about blue collars. A junior engineer with a good AI coding tool becomes productive in days on work that used to take months before anyone would trust them with it. The tool supplies the precise next step the way the app supplies Daniel's next turn, and the output is real, shippable, and often excellent. An operator who refused that gain on principle would be out of business by the end of the year, outrun by competitors who took it. The precision is not the villain of this chapter. It never is. The villain, if there is one, is the assumption that because the precision is good, more of it is always better, and that nothing is being spent to buy it.

## The Erosion

Now the other face, produced by the very same precision.

The more exactly a system specifies what to do, the less the worker ever practices the work of deciding what to do. In the same body of research, the precision that lifted accuracy by twelve percent cut exploratory behavior, the small self-directed experiments through which people discover better methods and build their own maps, by around thirty percent. Perceived freedom of action began to fall after as few as five prompts on a single task. The driver who is told every turn stops building the mental model of the city, exactly as the navigation users in Nottingham stopped learning the streets, except now there is a paycheck and a performance score riding on the following, which makes the following even harder to resist. The capability does not announce its own departure. It simply is not there on the morning the bridge is out.

Here is the twist that makes this a paradox and not a simple loss, and it is the most important sentence in the chapter. Self-efficacy can rise while autonomy falls. Daniel did not feel diminished. He felt good. He hit his targets, the app confirmed several times an hour that he was performing well, and he believed it, correctly, on every normal day. His confidence grew at the same time his actual independent capability shrank, and the two moved in opposite directions without him noticing, because nothing in his experience told him to look. That gap, between feeling capable and being capable, is the single most dangerous thing the system produces, because no one standing inside the gap has any reason to sound an alarm. The numbers are good. The worker is happy. The capability is quietly gone.

Picture the two curves on a single chart, because Marcus eventually did, and it is the chart that finally moved him. One line, self-efficacy, how capable Daniel feels, climbs steadily from his first week. The other line, his actual independent capability, the part that does not depend on the system, declines just as steadily over the same stretch. Early on the lines sit close together and nobody notices the divergence. By year three the gap between them is enormous, a driver who has never felt more sure of himself sitting at a barricade he has no idea how to get around. The danger of a system is not where it makes people visibly worse. We catch that. The danger is where it makes people feel better while making them more fragile, because feeling better is the one signal that guarantees no one will look closer.

## The Map in the Head

The thing Rosa had and Daniel lacked has a name worth dwelling on. She had a model in her head.

A model is not a memorized list of routes. It is an internal representation of how the city fits together, dense enough that when one road closes she can generate a new path she has never driven, because she understands the structure and not merely the steps. This is what the Nottingham drivers lost when the satellite voice did their thinking for them, and it is what every precision system quietly trades away. The system gives you the steps, instantly and reliably, and in exchange you stop building the model, because the model is expensive to build and the steps are free. The trouble is that steps without a model cannot adapt. The moment the situation leaves the territory the steps were written for, a person with only steps is stranded, while a person with a model improvises. Autonomy, stripped of its soft reputation, is exactly this: the possession of a model good enough to act without instructions.

And here is the cruelty of it. The model is invisible. You cannot see, from the outside or from any dashboard, whether a worker is operating from a rich internal model or merely executing the system's steps, because on every normal day the two produce identical behavior and identical numbers. You discover which one you have been growing only on the abnormal day, and by then it is far too late to grow the other. The whole discipline of leading through this tension is learning to invest in the invisible model during the long stretches when the visible steps are working fine, which runs against every instinct a metrics-driven operation has.

## The Threshold

Here the evidence gets specific enough to act on, and the specificity is good news, because it means the loss is not a fixed tax you pay for using these systems.

Two variables govern most of the damage. How fast the system speaks, and how ready the worker is for it. Where dashboards refresh about every five minutes and the workforce is not ready for the technology, meaning low on what researchers call technology readiness, a straightforward measure of comfort and fluency with new tools, perceived autonomy falls by roughly four tenths of a standard deviation. In the world of field effects, that is a large drop. It is the difference between a workforce that still thinks for itself and one that has handed the thinking over.

But raise the readiness and the same five-minute cadence does far less harm. Workers who are genuinely fluent and comfortable with the tools treat the prompts as information to weigh rather than orders to obey. They keep their own judgment switched on alongside the system, using it the way Rosa would have used a good map, as one more input rather than as the driver. Where readiness is high, the autonomy loss largely disappears even at the fast cadence. The competence gain, meanwhile, the rise of about three tenths of a standard deviation in self-efficacy, shows up either way. You get the confidence regardless. Whether you also get the erosion depends on conditions you can change.

Read that twice, because it contains the whole strategy of the chapter. The harm is roughly the product of cadence times unreadiness. You cannot always slow the cadence, because the cadence is often where the performance lives, and an operator who gives it up gives up the gains that justified the system. But you can always raise readiness. The lever the evidence puts in your hand is not use the system less. It is make your people more ready for it, and raising readiness is a thing leaders already know how to do. It is training, exposure, and the kind of fluency that lets a person argue with a tool instead of surrendering to it.

## The Cadence Trap

If readiness is the variable you can raise, cadence is the variable you will be tempted, constantly, to raise in the wrong direction.

Cadence is seductive because its gains are immediate and visible while its costs are delayed and hidden, which by now you will recognize as the signature of the whole paradox. Speed up the prompts, tighten the refresh, and throughput ticks up this week, on this dashboard, in front of this audience. The autonomy cost does not appear on any report for months, and when it does it wears another name. So the rational-seeming move, every quarter, is to crank the cadence a little higher, because the thing you can see improves and the thing you cannot see is, by definition, not yet visible. This is a ratchet. It turns only one way, toward faster, because nothing in the measurement system ever argues for slower.

The way out of a ratchet is not to fight it prompt by prompt. It is to put something on the other side of the scale that is as visible as throughput. If the only number on the wall is performance, cadence will rise forever. If there is also a number for autonomy, for voice, for the readiness that buffers the harm, then the conversation has two sides and the ratchet has a counterweight. This is why the very first move, back in Chapter 2, was to assign each tension an owner and a leading indicator. You cannot manage a paradox you can only see one half of, and the cadence trap is simply what happens to operators who never built the second gauge.

## Why Autonomy Is Not a Luxury

Some operators will hear the word autonomy and file it under soft, a comfort for people who do not have numbers to hit. That is a mistake, and it is an expensive one, so here is the hard-nosed case.

Autonomy is a basic psychological need, and when it is starved over time, motivation curdles into mere compliance. People stop bringing initiative they are no longer asked for. They either leave, like Carmen, taking the unit's retention down twenty-two percent on their way out, or they stay and quietly stop trying, which is more expensive and far harder to see. But the operator's case does not even need the motivation argument. Autonomy is not only a feeling. It is a capability, the muscle of independent judgment, and the day you need that muscle is precisely the day the system cannot help you, because if the system could have handled the situation you would not have needed the judgment in the first place.

So the bottom line for a hardheaded operator is this. You are not protecting autonomy to make your people happy, although it does that too. You are protecting it because it is the reserve capacity of your entire operation, the thing that keeps working on the day the thing that usually works has failed. A fleet, a ward, or a codebase with no independent judgment left in it is an operation running with no reserve. It will look magnificent on the dashboard right up until the morning the bridge is out, and then it will discover, all at once and at the worst possible time, exactly how much judgment it traded away for the last few points of efficiency.

This reserve-capacity logic is why the tension generalizes so cleanly across Grace's and Priya's worlds. A hospital that has trained its nurses to defer to the screen has spent its clinical reserve, and it will run beautifully until the case the screen misreads, which is the case that hurts someone. A software team that has let its engineers forget how to evaluate code has spent its engineering reserve, and it will ship smoothly until the failure the tool did not anticipate, which is the failure that takes down production. Different reserves, identical accounting. Every point of judgment you let atrophy is a withdrawal from an account whose balance you cannot read until the day you are overdrawn.

Marcus did the sum once, on the back of a meeting agenda. The efficiency the system bought him was a known, bankable number, printed every week. The judgment it was spending was unmeasured, which had quietly let him treat it as free. Nothing is free. It was simply a cost he had not yet been billed for, and the bill, when it finally came, would not arrive labeled as a routing decision. It would arrive as an accident, or a resignation, or a customer lost on the one day the system was wrong.

## What Marcus Tried First

Marcus's first instinct was the natural one, and it was wrong. He slowed the system down.

He cut the refresh rate, loosened the targets, and told the routing app to prompt less often. Performance dipped, which he had expected. What he had not expected was that autonomy did not come back. The drivers did not suddenly begin exercising judgment simply because the system nagged them less. The capability had not been waiting, intact, behind the prompts, ready to switch back on the moment the prompts eased. It had withered from disuse, and a withered capability has to be rebuilt, deliberately and over time. Slowing the system gave his drivers back some minutes. It did not give them back the city. He had treated a development problem as if it were a dosage problem, and the two require completely different cures.

The reframe that finally worked is the heart of the answer to this tension, and it sounds like a paradox of its own until you sit with it. The answer to too much specificity is not less specificity. It is shared authorship of the specificity. Bring the drivers into deciding which targets are right, which prompts genuinely help and which ones insult judgment they actually have, what the system should decide and what it should leave to the person. Keep the precision, because the precision works. Change who owns it. That single move turns a target from an order into an agreement, and a person can exercise judgment about an agreement in a way they never can about an order. We will build it into a full practice in the chapter on goal calibration. For now it is enough to see the shape of it, because the same shape, keep the gain and change the ownership, will reappear in every one of the four tensions.

There was a second thing Marcus learned from the failure of slowing the system down, and it changed how he thought about every tool in his network. He had assumed the autonomy lived in the gaps between the prompts, that if he simply made more gaps the judgment would flow back into them. But judgment is not a gas that expands to fill the available space. It is a skill that grows only when it is used, and the prompts had arranged things so that it was never used. More empty space did not summon the skill, any more than a longer silence summons a language you never learned. The drivers needed not just room to judge but reasons to judge, occasions that required it, and a system that treated their judgment as something it wanted rather than something it tolerated. That is a far larger redesign than a slider on a refresh rate.

He started small, because that is the only way these things start. He pulled six drivers, his most experienced and his newest mixed together, into a room once a month and put the routing rules on the screen, not as a finished product to accept but as a draft to argue with. Which prompts actually helped? Which ones interrupted a judgment the driver had already made correctly? Where should the system defer to the person, and where should the person defer to the system? The first meeting was awkward and mostly silent, because the drivers had been trained for years that the system was not theirs to question. By the third meeting they were arguing, which was the entire point. A driver who argues about a target has to exercise judgment about it, and a driver exercising judgment is a driver rebuilding the very muscle the system had let waste. The targets barely changed. The drivers changed completely.

The principle to carry out of Marcus's tension is short. The clearer the instruction, the more deliberately you must protect the judgment it replaces, because the instruction will not protect it for you, and the worker following the instruction will not even notice it is gone.

Notice that the principle does not ask you to choose between precision and judgment, which is the false choice that traps most operators. It asks you to do two things at once, to deploy the precision and to protect the judgment, holding both at full strength, which is the dynamic equilibrium from Chapter 2 made concrete for the first time. The operator who hears only the first half builds a fleet of Daniels. The operator who hears only the second half, who rips out the precision in the name of autonomy, gives up the gains and, as Marcus learned, does not even get the autonomy back. The skill, and it is a skill, is to keep both hands working. Precise targets in one hand. Deliberately protected judgment in the other. The chapters in Part Three are the training for that second hand, the one most operations have allowed to go weak.

Marcus put a single question at the top of every operational review after that, and it is a good one to steal. Not only how are the numbers, but if this system were wrong tomorrow, what would my people be able to do? An operation that cannot answer that question has not yet measured its most important risk.

Marcus's tension is about the precision of the target. The one we turn to next belongs to Grace, and it is about the speed of the feedback, and it does something stranger and more dangerous than eroding autonomy. It does not make people less capable. It makes capable people go quiet.

### THE EVIDENCE

Specificity and autonomy is the most common of the four tensions, present in fifty-eight percent of the reviewed studies. Precise, fast-refreshing targets raised task accuracy by about twelve percent and reduced exploratory behavior by about thirty percent, with perceived freedom dropping after roughly five prompts on a task.

The first proposition: at a five-minute refresh with low technology readiness, perceived autonomy falls about 0.40 SD while self-efficacy rises about 0.30 SD; at high readiness the autonomy loss largely disappears at the same cadence. The harm is cadence times unreadiness, and readiness is the variable you control.

### MONDAY MORNING

Find your Daniels and your Rosas. Identify one team on a fast-refreshing system, and ask two questions. How often does the system prompt them in an hour, and how fluent are they really with the tool, not how trained but how willing to argue with it. Then ask the question that matters most: if the system were wrong for a day, who on this team could still do the job, and who would be sitting at the barricade?

CHAPTER FIVE

# Cadence and Safety

*Watch people closely enough, and you will see everything except the thing they have decided not to show you.*

The patient in bed nine was going to be fine, but Ada did not know that yet, and the screen was telling her not to worry.

Ada had been a charge nurse for nineteen years, long enough that she could feel a ward the way an old sailor feels weather coming. Bed nine was a woman two days past surgery, and every number on her was inside the lines. Her vitals were good. The early-warning score the system computed for her was green, which meant stable, which meant move along. And yet. The color in the woman's face was a shade Ada did not like. She was restless in a way that did not match her chart. Her daughter sat at the bedside and kept catching Ada's eye across the room. None of it was measurable. All of it, to Ada, was a sentence she had read a thousand times before.

Two years earlier Ada would have acted on it without a moment's hesitation. She would have walked over, looked closer, and called the resident with the sentence every good nurse learns to say: I cannot point you to a number, but I do not like how she looks. That sentence has saved more lives than most of the equipment on the floor. It is the sound of human judgment reaching past the data to the thing the data has not caught yet.

Ada did not say it. She stood at the station and watched the screen, which said stable, and she waited. Here is why she waited, and it is the whole of this chapter. Eight months earlier the unit had gone live with a real-time dashboard that scored every nurse on response times, task completion, and protocol adherence, refreshing every couple of minutes and visible to everyone, the nurses themselves and the managers above them. Escalating a patient the system had rated stable, and turning out to be wrong, was logged as a false alarm. False alarms pulled down your adherence score. A low adherence score meant a conversation with someone holding a printout. No one had ever told Ada to wait for the screen. She had simply learned, the way water learns the shape of a pipe, that it was safer to wait until the screen agreed with her before she acted on what she already knew.

Bed nine deteriorated about forty minutes later. The score finally turned, Ada moved fast, the rapid response team arrived, and the patient was stabilized with no lasting harm. It was a near miss, the kind that never becomes a report because nothing technically went wrong. But Grace reviewed her units' incidents personally, and she saw the forty-minute gap between the moment Ada first felt it and the moment the screen finally let her act, and she asked Ada about the gap.

"The screen said stable," Ada told her. "And if I had called and been wrong, that is a mark on me. I have learned to wait for the screen to agree with me." Grace told me that sentence kept her up at night. She had installed a system to make her hospitals safer, and it had taught one of her finest nurses to silence the exact instinct that had made her one of her finest nurses.

Grace did something most executives never do. She went and sat with Ada for an hour, not to reprimand her but to understand her, and what she found was not a nurse who had gotten worse. It was a nurse who had gotten careful in the one direction that hurts patients. Ada could still feel bed nine going off as surely as she ever had. The instinct was perfectly intact. What had changed was the cost of acting on it, and Ada had simply done the math, over and over, until the math became a habit and the habit became silence. The system had not damaged her judgment. It had taught her to keep it to herself, which for a charge nurse is very nearly the same thing.

## The Mastery Signal

Begin where we always begin, with what the cadence does well, because it does a great deal well, and Grace would never have installed it otherwise.

Rapid feedback is one of the most powerful forces in all of human learning. It confirms progress, it removes the corrosive uncertainty of not knowing where you stand, and it lets a person correct course before a small error becomes a habit. A new nurse on Grace's units learned the rhythms of the floor faster than any cohort before her, because the system told her in near real time whether she was keeping pace. Coordination improved across the whole ward. Handoffs tightened. Everyone could see the same picture at the same moment, which in a hospital is worth a great deal. In the first months the nurses even reported that the system felt fairer than the old way, where recognition had depended on whether the right manager happened to be looking in your direction at the right time.

Call this the Mastery Signal. Fast, frequent feedback tells a person that they are getting good, and the felt experience of visibly getting good is among the most motivating things a human being can encounter at work. None of it is an illusion, and the cadence is no more the enemy of this chapter than precision was the enemy of the last. The trouble, as always, is that the same feature that delivers the gift also delivers something else, through the very same mechanism, at the very same time.

## The Spotlight Threat

Now the same feature, the constant visibility, seen from its other side.

When every action you take is scored in near real time and visible to everyone, your work quietly becomes a stage. You do not decide to perform. You simply begin, without noticing, to manage how you look at least as much as what you do, because looking is now a measured output with consequences attached. Call this the Spotlight Threat. The same act, making performance continuously visible, that produces the Mastery Signal also produces the Spotlight Threat, because being seen and being watched are the same event felt from two different emotional directions. One says you are getting good and someone notices. The other says you are being judged and you had better not slip.

The first casualty of the spotlight is candor. Across the reviewed studies, once feedback passed roughly thirty prompts an hour, the rate at which people raised concerns, asked uncertain questions, or admitted they were not sure dropped by about a quarter. The people did not become less competent. They became less willing to be seen being uncertain, and uncertainty voiced early is precisely how problems get caught before they grow. A nurse who waits for the screen, an engineer who does not flag the code that looks wrong, a worker who swallows the question rather than step into the light to ask it. The spotlight does not make people worse at their jobs. It makes them quieter, and quiet is far more dangerous than bad, because bad announces itself and quiet does not.

Notice the cruel symmetry with the previous chapter. Precision made Daniel feel capable while making him fragile. Cadence makes Ada look compliant while making her silent. In both cases the visible signal, confidence in one, compliance in the other, moves in the opposite direction from the thing that actually keeps an operation safe, which is judgment in one case and candor in the other. The system is optimizing the appearance and starving the substance, and it cannot tell the difference, because from the outside, on every instrument the organization owns, the appearance and the substance look exactly alike.

## The Two Audiences

There is a simple way to see why the spotlight does what it does. When work is scored live, the worker is performing for two audiences at once.

The first audience is the real one: the patient, the customer, the code, the actual work the job exists to do. The second audience is the screen, and behind the screen the manager, and behind the manager the consequence. On a good day the two audiences want the same thing and there is no conflict. The trouble arrives on the days they diverge, the days when what the patient needs and what the screen rewards point in different directions, because those are exactly the days that matter most. Ada's patient needed her to escalate. Ada's screen rewarded her for waiting.

When the two audiences disagree, the worker has to choose which one to serve, and the screen holds a structural advantage in that choice, because the screen keeps the tally and the patient does not. You will serve the audience that grades you, eventually, no matter how much you care about the other one, because caring does not sign your review and the screen does.

This is why the intensity of the cadence matters so much. A quiet system, checked now and then, leaves the worker mostly facing the real audience, glancing at the screen occasionally. A system refreshing every two minutes puts the screen in the worker's face continuously, until the second audience crowds out the first and the worker is, in practice, working for the watcher. Nobody decides to make that switch. It is the slow gravity of constant measurement, and it pulls in one direction only.

## Surveillance Is Not Feedback

Underneath all of this sits a confusion that almost every organization makes, and naming it is half the cure. Feedback and surveillance are not the same thing, even when they ride on identical data.

Feedback is information delivered to the person doing the work, for the purpose of helping them do it better. Surveillance is information delivered to someone watching the work, for the purpose of judging it. The very same data stream, response times and error rates and task completion, can be either one, and what decides which it is has nothing to do with the data and everything to do with who it is for and what it triggers. The instant a number that was meant to help a nurse learn becomes the number that decides whether she has a difficult conversation with her manager, it has stopped being feedback and become surveillance, and her nervous system knows the difference even when the org chart does not.

The deep error in Grace's original system, and in most systems like it, was to run both functions through one screen. The same dashboard coached and policed, which meant every helpful signal arrived wearing the uniform of a threat. You cannot learn much from a teacher who is also your judge and is grading you in real time, because you spend your attention managing the grade instead of absorbing the lesson. The single most powerful structural move available to a leader here is also the simplest to state and among the hardest to do politically. Separate the two. Let the learning signal be a learning signal, owned by the worker, and put the evaluation on a different clock and a different screen. The data can be the same. The wiring must not be.

## Psychological Safety, Defined for Operators

The name for what Ada lost is psychological safety, and the phrase has a soft sound that does it a real disservice, so let me define it the way a hardheaded operator should hear it.

Psychological safety is not comfort, and it is not the absence of pressure or accountability. It is the shared belief that a person can take an interpersonal risk, ask the obvious question, admit the mistake, raise the concern that might turn out to be nothing, without being punished for having done so. Research on teams has found, again and again, that the highest-performing teams are not the ones that report the fewest errors. They are often the ones that report the most, because they are safe enough to surface them, which is exactly how the errors get caught and fixed before they compound. A team that reports very few errors is usually not making fewer errors. It is hiding more of them, and a hidden error is just an error that has been given time to grow.

Every organization runs up a balance of what you might call speak-up debt, the accumulated total of concerns that occurred to someone and were never voiced. The debt is invisible, it produces no statement, and it charges no interest until the day it is called, all at once, in the form of an incident that everyone afterward turns out to have half seen coming. The most dangerous number in any operation is the one that does not exist, the count of things people noticed and decided not to say. A high-cadence, high-spotlight system is a machine for growing that number, because every increment of visibility raises the felt cost of being the person who spoke up and was wrong.

In a hospital this is not a management abstraction. The unraised concern is the near miss that becomes the miss. The entire safety architecture of modern medicine rests on the willingness of the person closest to the patient to say something is wrong here before the numbers prove it, because by the time the numbers prove it the window to act has often closed. A system that quietly raises the personal cost of speaking up has not made the ward safer. It has disarmed the ward's single most important safety mechanism while making the safety dashboard look better than it ever has. That is the worst of all possible outcomes, a real decline in safety wearing the costume of an improvement, and it is exactly what Grace had built without meaning to.

## The Anatomy of a Swallowed Concern

It is worth slowing down the three seconds in which a concern gets swallowed, because the whole tension lives inside them.

A nurse notices something. In the next breath, before she is even aware of deliberating, a calculation runs. If I speak and I am right, I prevent a problem nobody will ever know was coming, so there is no visible reward. If I speak and I am wrong, the system logs a false alarm, my score dips, and I may have to explain myself. If I stay quiet and I am wrong, the problem surfaces later as someone else's catch, or as a bad outcome that gets attributed to the patient's condition rather than to my silence. The costs and rewards are wildly asymmetric, and every one of them points the same way, toward saying nothing. The astonishing thing is not that nurses go quiet. It is that, under those incentives, any of them still speak.

Do that calculation a few hundred times and it stops being a calculation. It becomes a reflex, and then a personality. The nurse does not experience herself as suppressing concerns. She experiences herself as a careful professional who waits for confirmation, which sounds like a virtue and is, in fact, the precise shape of the damage. This is how the spotlight does its worst work. Not by frightening people in the moment, but by slowly editing the kind of professional a person understands themselves to be, until the silence feels like good judgment and the speaking up feels like recklessness, and the whole safety culture has quietly inverted while every dashboard stayed green.

## The Threshold

Here is the proposition, and like the last one it carries a condition you can act on, which is the part that turns a warning into a plan.

Where feedback runs hot, past roughly thirty prompts an hour, and the system is opaque, meaning people do not understand why it flags what it flags or scores what it scores, psychological safety falls by about half a standard deviation. That is a large effect, large enough to change the behavior of an entire unit, and the dissent that early problem-catching depends on drops by about a quarter alongside it. A floor that loud and that opaque will go quiet, reliably, and the quiet will read as calm.

But the opacity is half of that equation, and it is the half you can change without giving up the cadence. When people understand the logic of the system, when a nurse knows why a patient is scored the way they are and why her own actions are flagged the way they are, the spotlight stops feeling like surveillance and starts feeling like information. The harm is roughly cadence times opacity. You cannot always slow the cadence, because the speed is often where the coordination gains live. But you can almost always reduce the opacity, and reducing it takes much of the threat out of the spotlight even when the cadence stays high. Information that you understand is a tool. The identical information, delivered as a verdict you cannot question, is a threat. Same data, opposite effect, and the difference is whether the box is open.

Hold that thought, because it points straight at the fourth tension, the one between transparency and trust, which we reach in two chapters. For now the operator's takeaway is narrow and useful. If you are going to watch people closely, you owe them a clear understanding of what you are watching for, or the watching will quietly cost you the candor you most need on the day it matters.

## Why the Quiet Is So Dangerous

Of all four tensions, this is the one that hides the best, and it hides the best because its symptom is an absence.

When autonomy erodes, you eventually get turnover, and turnover is at least a number. Carmen showed up in the data late and mislabeled, but she showed up. When candor erodes, you get nothing at all. Silence leaves no trace. The concern that was never raised does not appear on any report, does not move any metric, does not trigger any review. It is the only failure mode whose entire signature is the absence of a signal, which leads to a genuinely unsettling conclusion. The better your safety dashboards look, in some cases, the more worried you should be, because a perfectly quiet ward is indistinguishable, on every instrument you own, from a ward where everyone has simply stopped speaking. You cannot tell health from silence by looking at a screen, because both of them look like green.

And it is not only hospitals. Priya saw the same quiet creeping into her standups, where junior engineers had stopped flagging the AI-generated code that looked wrong to them, because raising it slowed the sprint and the velocity dashboard was watching. Marcus saw it on his floors, where workers had stopped reporting a broken process, because reporting it meant stepping off the timed task and into the light. The same mechanism, the same silence, the same dangerous calm, in three industries that share nothing but the shape of their management. Wherever the spotlight burns hot enough, people learn to perform stillness, and an organization full of people performing stillness is an organization that has gone blind to its own early warnings.

## What Grace Tried

Grace's first instinct was the one every leader reaches for, and it does not work. She told people to speak up.

She ran the sessions. She put up the posters. She gave the heartfelt talk about how every voice mattered and no concern was ever too small. And nothing changed, because you cannot exhort safety into existence while the system quietly punishes the very behavior you are exhorting. People do not respond to the talk. They respond to the incentives, and the incentive structure was saying, in a language far louder than any poster, that speaking up before the screen agreed with you was a risk to your score. When the words on the wall and the math in the system disagree, the system wins every time, because the system is the one keeping the tally.

What worked was structural, and it came in three moves we will build out in full later. First, where she could, she slowed the cadence, widening the refresh on the metrics that did not actually need to be instant, because not everything has to be measured every two minutes and most things do not. Second, she added transparency, so that nurses understood what the early-warning system was truly watching and why, which let them argue with it instead of fearing it. Third, and most important, she separated the signals. The dashboard that helped a nurse learn was pulled apart from the dashboard managers used to evaluate her, because the moment the learning tool and the surveillance tool are the same screen, every piece of feedback arrives carrying a threat, and very little learning survives under threat.

She also built a protected pause, a short and regular and deliberately off-the-record huddle where nurses could raise the I-cannot-point-to-a-number concerns without any of it touching a score. That huddle is the seed of the hybrid coaching we reach in Part Three. The point of every one of these moves was the same. Keep the gift of the cadence, the fast signal that genuinely helps people learn and coordinate, and take the threat out of it, so that being seen stops meaning being judged. Ada did not need less feedback. She needed feedback that did not punish her for the most valuable thing she did, which was to notice, before anyone else could, that something was wrong.

The results did not show up on the safety dashboard, which was rather the point, because the safety dashboard had been part of the problem. They showed up in the huddles, where the I-do-not-like-how-she-looks concerns started coming back, tentatively at first and then freely, once the nurses believed that raising one would cost them nothing. Grace began counting those concerns, not to score anyone but to watch the number climb, because a rising count of voiced concerns, most of which turn out to be nothing, is the single best leading indicator of a unit that is safe to speak in. She had found a way to measure the very thing whose absence had nearly cost a patient, by measuring the willingness to speak rather than the silence, and willingness, unlike silence, leaves a trace you can count.

The principle to carry out of Grace's tension is this. If you are going to make work visible, you must make it safe to be seen, because visibility without safety does not produce better work. It produces better-looking work, performed by people who have quietly stopped telling you the truth.

Ada's tension is about the speed of the feedback. The next one belongs to Priya, and it is about the price of speed itself. Her engineers were not going quiet. They were going fast, faster than anyone in the company's history, and the faster they went the less they learned, until she looked up one day and realized she was putting up a taller and taller building on a thinner and thinner foundation.

### THE EVIDENCE

Cadence against psychological safety appeared in forty-seven percent of the reviewed studies. Rapid feedback validates and coordinates, the Mastery Signal, while the same constant visibility creates the Spotlight Threat. Past roughly thirty feedback prompts an hour, the willingness to raise concerns fell by about twenty-five percent.

The proposition: where cadence runs past about thirty prompts an hour and the system is opaque, psychological safety falls roughly 0.50 SD. Transparency is the moderator. Understood feedback reads as information; the same feedback as an unexplained verdict reads as a threat. You may not be able to slow the cadence, but you can almost always open the box.

### MONDAY MORNING

Take one team on a high-cadence system and ask: in the last month, who raised a concern that turned out to be nothing? If you cannot name anyone, that is not because there were no concerns. It is because the spotlight is working exactly as designed. Then check one thing more. Does the screen that coaches your people also discipline them? If it is the same screen, you have wired a threat into every piece of feedback, and your people have noticed even if you have not.

CHAPTER SIX

# Efficiency and Learning

*Efficiency removes everything unnecessary. Learning is mostly what happens in the part you removed.*

The payment service went down at two in the morning, and the only person who truly understood it was on a plane.

Kofi was on call. He was twenty-six, eighteen months at the company, and by the velocity dashboard the most productive engineer on his team. He shipped more features, faster, than engineers twice his age, because he was fluent in the AI coding tools the way Daniel had been fluent in the routing app. The tool wrote, he reviewed it lightly, it shipped, and the dashboard glowed a satisfied green. By every number the company tracked, Kofi was the future and Kofi was winning.

At two in the morning the payment service began to fail, and customers could not check out, and the revenue graph started sliding toward the floor. Kofi opened the code, code the tool had largely written six months earlier, and he could not find the problem. Not because he was not smart. He was very smart. He could not find it because he had never built a model of how the payment system actually fit together. He knew how to ask the tool to produce a piece of it. He did not know how the pieces held each other up, which is an entirely different kind of knowledge and the only kind that helps at two in the morning. He asked the tool to find the bug. The tool, which also did not understand the architecture, confidently proposed three fixes. None of them worked. Two of them made it worse.

Lena was the senior engineer who had been there six years, who had built half the system by hand back before the tools existed. She was on leave. They paged her anyway. She logged in from her hotel room, read the symptoms for a few minutes, and found the bug in a place the tool would never have thought to look, because finding it required understanding why the system had been built the way it was, the history and the trade-offs and the load-bearing decisions that live in no file and no comment. She fixed it, went back to sleep, and the dashboard never recorded that the entire business had just been rescued by the one engineer it rated as less productive than Kofi.

Priya told me this the way Marcus had told me about the bridge, as a near miss that frightened her more than the things that had actually gone wrong, because she could see the future inside it. Lena was leaving in a year. And there were no new Lenas coming up behind her, because the system that produced Kofis so efficiently had stopped producing Lenas. "I am putting up a taller building every quarter," Priya said, "on a foundation that is getting thinner every quarter, and the maddening part is that the two trends have exactly the same cause."

The next morning at standup, Kofi was not chastened. He was impressed. "Lena is a genius," he told the team, and he meant it as the highest praise. Priya heard something else in it, something that worried her more than the outage had. Kofi did not think of Lena's skill as something he was supposed to be acquiring. He thought of it as a rare gift that some people simply have, like perfect pitch, and that the rest of them could safely route around by paging the person who had it. He did not see a ladder he was meant to climb. He saw a specialist he could call. And a team that has stopped seeing the ladder is a team that has stopped climbing it, which is perfectly fine, right up until the specialist retires.

## The Success Cue

Grant the gain first, and in this case the gain is enormous, because any operator who waves it away will be out-shipped within the year by everyone who did not.

Optimization systems lift output by removing friction, standardizing the best-known path, and steering people firmly down it. In software, AI coding tools raised measurable productivity by around eighteen percent in one careful study, and in many teams the felt gain is far larger than that. A feature that took a week takes a day. The tedious boilerplate simply vanishes. The tool supplies the next step, correct and instant, and the engineer moves on to the next thing while a competitor down the road is still typing. Call this the Success Cue. Every quick win the system delivers is a signal that says this way works, do more of this, and the signal is usually telling the truth. The work really does get done faster, and quite often it gets done better. An operator who feels no pull toward that gain is not principled. They are about to be unemployed.

## Practice Crowding

Now the other face, produced by the very same removal of friction.

The friction the system removes is not all waste. Some of it is practice. The afternoon spent writing a function by hand, the slow read through an unfamiliar corner of the codebase, the exploratory debugging that goes nowhere three times before it teaches you how the system actually breaks. These are inefficient by any measure the dashboard understands. They are also precisely how a junior becomes a senior. The Success Cue says skip them, the tool can handle that part, and on any single day skipping them is the correct call. Across a thousand single days, skipping them is how you assemble a team of people who can produce code and cannot understand it. Call this Practice Crowding. The same act, eliminating slack, that produces the Success Cue also crowds out the deliberate practice through which deep skill is built. Across the studies, past about twenty optimization nudges in a shift, exploratory behavior fell by around fifteen percent.

It is the Nottingham drivers one final time, because the pattern is that fundamental. The satellite voice got them there and stopped them learning the city. The coding tool ships the feature and stops the engineer learning the system. The scheduling system smooths the ward and stops the nurse building her own model of the floor. The efficiency is real and the learning loss is real and they are the same event, not two separate outcomes to be weighed against each other on a balance. You are not trading some learning for some efficiency. You are buying efficiency with learning, in a single transaction, often without seeing the price tag.

## What the Tool Cannot Teach

There is a difference between having an answer and being able to generate one, and it is the difference that decides everything here.

When the tool hands Kofi a working function, it transfers an answer. What it cannot transfer is the understanding that would let him produce that answer himself, or recognize when the answer is wrong, or build a different one when this one does not fit. Understanding is not a thing that can be handed across. It is a residue, the stuff left in a person after they have struggled with a problem and come out the other side, and the struggle is not incidental to it. The struggle is how it is made. This is why you cannot learn to swim by watching, or to debug by reading the fix. The flailing is the lesson, and a tool that removes the flailing removes the lesson along with it, delivering the destination while quietly canceling the journey that would have turned you into someone who could find the destination alone.

Educators have a name for the narrow band where real learning happens, the zone just past what you can already do, where a task is hard enough to demand effort but not so hard that you simply fail. Growth lives in that band and almost nowhere else. The trouble with a highly capable tool is that it collapses the band. It takes the tasks that would have sat in a junior's growth zone, the ones just beyond their current reach, and completes them instantly, so the junior never spends time in the only place they would have grown. They operate, all day, comfortably inside what they already know, handing everything harder to the tool, and comfort sustained long enough is just another word for stagnation.

Kofi was the proof. He had shipped more code than anyone, and almost none of it had ever sat in his growth zone, because the moment a task got hard enough to teach him something he handed it to the tool, which is the rational move on every individual day and a quiet catastrophe across two years of them. He had eighteen months of tenure and perhaps three months of actual learning, the rest having been performed by the tool while he supervised. His resume said eighteen months. His judgment said three. No interview, and certainly no dashboard, was built to tell the two apart.

## The Reinforcing Loop

This tension has one feature the others do not, and it is what makes it the most dangerous over the long run. It accelerates itself.

Walk the loop once. The tool delivers a quick win. The quick win builds the engineer's confidence in the tool's way of doing things. That confidence reduces the felt need to explore alternatives or to understand what sits underneath, because the tool's way is plainly working. Less exploration means less learning. Less learning means the engineer is less equipped to evaluate the tool's output, which makes them rely on the tool more, which produces more quick wins, which deepens the confidence, and around it goes. Each turn of the loop raises output and lowers understanding, and because output is measured and understanding is not, every single turn looks like progress on every screen the company watches. It is a flywheel that spins toward fragility, and it spins faster the better the tool gets.

And here the loop does something genuinely cruel, the same thing Daniel's app did to him. It inverts the relationship between confidence and competence. Kofi felt more expert every quarter, and he was right to, in the narrow sense that he could produce more, faster, than ever before. His confidence was real and it was rising. His independent competence, the part that does not depend on the tool, was falling over the same period. The two moved in opposite directions, exactly as they had for Daniel, and for the same reason. The system supplies the experience of mastery while removing the practice that mastery is actually made of. A junior who feels like an expert and cannot debug the payment system at two in the morning is not lying about how he feels. He has been given the feeling without the substance, which is the most dangerous gift a tool can offer, because the feeling is precisely the thing that stops him from going looking for the substance.

This is why a thinner foundation every quarter was the exactly right way for Priya to say it. The gap between what the team can produce and what it can understand widens silently, turn after turn, and it stays invisible until the night the building is forced to stand on its foundation alone. Which is to say, until the night the tool is wrong and the one senior who could have caught it is on a plane. The loop does not feel like decline while you are inside it. It feels like the best quarter you have ever had, right up until it does not.

## The Two Ledgers, Again

Marcus had his two ledgers, and Priya had hers, and they were the same two ledgers in different currencies.

On the visible ledger, the one the company reviewed every week, Priya's team had never looked better. Velocity up, cycle time down, features shipping at a rate that made the roadmap look conservative. Every number that had a dashboard was pointing the right way. On the invisible ledger, the one no system kept, the entries ran the other direction. The number of engineers who understood the payment system end to end had dropped from five to two over eighteen months. The share of production code that no living person fully understood was climbing. The depth of the bench, the count of people who could handle a genuine surprise, was thinning even as the headcount grew. None of it had a column, so officially none of it was happening, which is precisely the sentence Grace had used about her wards and Marcus had used about his floors.

The foundation has no dashboard. That is the whole problem in five words. An organization will defend, optimize, and celebrate whatever it measures, and it will spend, silently and without resistance, whatever it does not. Understanding is unmeasured, so understanding is what gets spent, and it gets spent first and fastest precisely because spending it shows up nowhere until the building begins to lean.

## The Threshold

The proposition, with its condition, which as always is the hopeful part of the story.

Where optimization nudges run heavy, past about twenty in a shift, and the workforce has a low learning orientation, exploratory behavior falls by about fifteen percent and keeps on falling. Learning orientation is the disposition, in a person or in a culture, to treat work as a chance to get better rather than only as a chance to perform well today. It can be measured, at least roughly, and where it runs high, people preserve their exploration even under heavy nudging, because they genuinely want to understand and not merely to ship. They argue with the tool. They read the code the tool wrote. They ask why it works, not just whether it works.

So the harm is roughly nudges times low learning orientation, and learning orientation is something a leader can build. You cannot always cut the nudges, because the nudges are where the throughput lives and the competition is using every one of them. But you can raise the learning orientation of your teams, and you can protect, deliberately, the practice the nudges would otherwise crowd out. The lever the evidence puts in your hand is not use the tool less. It is build people who use the tool the way Lena would rather than the way Kofi does, and that is a thing a leader can set out to do on purpose, with structure, on a schedule. We will build it into a full practice in the chapter on rotational upskilling.

## The Generational Time Bomb

There is a longer and far more frightening version of this tension, and Priya had already seen the shape of it.

The seniors who can supervise the tools, the Lenas, built their judgment the hard way, before the tools existed. They spent years doing the inefficient work, and that work is exactly what made them into people who know what good looks like, which is exactly what you need in order to catch a confident tool being quietly wrong. The juniors are being built on a different plan. They are being built on top of the tools, and they are never doing the inefficient work that made the seniors, which means they are not acquiring the judgment that would one day let them supervise the tool in turn. The ladder that produced Lena has had its bottom rungs removed as an inefficiency, and no one decided to remove them. The optimization removed them, quietly, because they did not pay off this quarter.

Run that forward ten years. The Lenas retire. The Kofis are senior now by title and tenure, but they were never made into Lenas, because the slow apprenticeship that made Lenas was optimized away before they could climb it. So who reviews the tool's output then? Who finds the bug at two in the morning? The organization will have optimized away the very capability it most depends on, and it will not notice the loss until the capability is needed and absent, which is the signature of this entire book. This is Marcus's Rosa problem and Grace's veteran-nurse problem wearing a hoodie. The same hollowing of the deep bench, in three industries, from one cause. The bench looks full today. It is the bench ten years out that has quietly stopped being stocked.

## Why Seniors Stop Teaching

There is a second engine driving the bench toward empty, and it is quieter than the first. The tools do not only stop juniors from learning. They stop seniors from teaching.

For as long as there have been crafts, the senior taught the junior by reviewing the junior's work. The junior tried, the senior looked at the attempt, and in the gap between them, the corrections and the questions and the why-did-you-do-it-this-way, the craft passed from one generation to the next. Code review was the classroom, even when no one called it that. Now the tool does the junior's work, and review changes character. The senior is no longer looking at a person's attempt and teaching into it. They are looking at a tool's output and checking whether it runs, and that is not teaching, it is inspection. The channel through which judgment used to pass from old to young has not been improved. It has been unplugged, and because the work still ships and the dashboard still glows, nobody notices that the transmission has stopped.

Worse, the seniors are busy, and a tool's output is usually good enough to wave through, so review degrades into rubber-stamping, which feels efficient and is, in the way that mortgaging the future always feels efficient in the present. The senior stops investing the slow attention that teaching requires, the junior stops receiving it, and both of them are rewarded by the dashboard for the disinvestment. The apprenticeship does not end with a decision. It erodes, review by hurried review, until one day there is no one left who learned the craft the way the last generation did, and no one who even remembers that the review used to be where the learning happened.

Priya tested this on herself. She pulled the last twenty code reviews she had personally signed off on and asked how many had involved her actually teaching anything, as opposed to confirming that a tool's output compiled and passed its tests. The honest answer was two. Eighteen times she had been an inspector. Twice she had been a mentor. She had not decided to stop teaching. She had simply been carried, review by review, into a role that looked like teaching, was paid like teaching, and transmitted almost nothing. And she was the best senior she had. If she had drifted that far without noticing, she had no reason to believe the rest of her seniors had not drifted further.

## What Priya Tried

Priya's first instinct was Marcus's instinct, and it failed in exactly the same way. She tried to slow it down.

She put limits on the tool. She mandated manual work for the juniors. She slowed the very thing that had made her team competitive. Velocity dropped, as she had expected. What she had not expected was the rest of it. The juniors were furious, two of her strongest left for companies that let them move fast, and, crucially, the ones who stayed did not actually get better at understanding the system. They simply shipped less. She had, exactly like Marcus, treated a development problem as if it were a dosage problem, and she learned the same hard lesson he had. You do not rebuild a capability by removing a tool. The capability has to be grown, on purpose, and growing it is a completely different project from limiting the tool. Removing the tool just gives you slower people who are no wiser.

What worked was to keep the efficiency and re-insert the practice on purpose. She rotated engineers, on a schedule, onto deliberately hard problems with the tools turned down, not as a punishment but as training, the way a pilot logs hours in the simulator even though the autopilot flies the actual plane. She turned code review from a gate into a classroom, where the question was no longer only is this correct but why is it built this way, and what would you do if it broke at two in the morning. She required juniors to explain the tool's code in their own words before it shipped, which turns out to be the fastest way ever devised to discover whether a person actually understands what they are about to put into production. None of it slowed the tool. All of it began, slowly, to rebuild the foundation. That deliberate rebuilding is the lever we reach in the chapter on rotational upskilling.

The simulator analogy did more work than Priya expected, because it reframed the whole thing for her engineers. No pilot believes that logging hours in the simulator is a vote of no confidence in the autopilot. It is simply understood that the autopilot flies the plane on every normal flight, and that the pilot trains relentlessly for the abnormal one, because the abnormal flight is the entire reason there is a human in the cockpit at all. Priya started saying exactly that to her teams. The tool ships the features. You train for the outage. Both are real parts of the job, and the second one does not appear on the velocity dashboard for the same reason the pilot's simulator hours do not appear on the flight's on-time record. It is not measured because it is insurance, and insurance only ever proves its worth on the day you needed it and were glad you had paid the premium.

The principle to carry out of Priya's tension is this. A tool that does your hardest thinking for you is a loan taken against your future competence, and the interest comes due on the day the tool is wrong. Take the loan, because the gains are real and refusing it is its own kind of bankruptcy. But pay it back deliberately, by practicing the very thing the tool does for you, or you will find yourself insolvent at exactly the moment you can least afford to be.

Priya put her own question at the top of her reviews after that, a cousin of the one Marcus asked. Not only how fast are we shipping, but how many people here could rebuild this if they had to. An operation that ships quickly and cannot rebuild what it ships has not measured its real exposure. It has only measured its speed toward the edge of it.

Efficiency and learning is the third tension, and it shares a hidden ingredient with the second. In both, the harm grew worse when people did not understand the system, when the tool or the dashboard was a black box they could not argue with. That ingredient, transparency, turns out to matter so much that it forms the fourth tension on its own. And it carries a genuine surprise, one that has undone many well-meaning leaders. Opening the black box, it turns out, is necessary, and nowhere close to sufficient.

### THE EVIDENCE

Efficiency against learning depth appeared in forty-four percent of the reviewed studies. Optimization tools raised productivity by roughly eighteen percent, the Success Cue, while the same removal of slack produced Practice Crowding, with exploratory behavior falling about fifteen percent past roughly twenty optimization nudges in a shift.

The proposition: heavy nudging combined with low learning orientation drives the exploration loss, and the loss compounds through a reinforcing loop of quick wins and shrinking understanding. Learning orientation is the moderator and the lever. You cannot always cut the nudges, but you can build teams that explore anyway, and protect the practice the nudges would crowd out.

### MONDAY MORNING

Find your Lena and your Kofi. Pick one recent piece of work the tool largely produced, and ask the engineer who shipped it to walk you through why it is built that way and what they would do if it failed in production at two in the morning. If they can answer, you are growing a Lena. If they can only tell you what the tool did, you are growing a beautiful, fragile building, and you have just measured, for free, the current thickness of your foundation.

CHAPTER SEVEN

# Transparency and Trust

*Explaining a decision to someone is not the same as giving them a say in it, and they can always tell the difference.*

Priya did everything right, and trust went down.

She had heard the lesson of the previous tension and taken it to heart. Her engineers did not trust the system that scored their code, the automated reviewer that flagged quality, assigned risk, and fed a number into the picture their managers saw. The obvious fix, the one every thoughtful leader reaches for, was transparency. So Priya opened the box, all the way. She published the model's logic, documented every metric and weight, held a town hall where she walked through exactly how a score was computed, and invited every question anyone had. She was proud of it, and she had a right to be. It was more genuine openness than most companies would ever offer their people.

Comprehension shot up. Within weeks her engineers could explain the scoring system better than the vendor who had built it. And trust did not move. In a couple of teams it dropped. Priya could not make sense of it. She had handed her people the one thing they had asked for, complete understanding, and they trusted the system less for having received it.

The sentence that finally explained it to her came from Lena, the senior engineer who had found the bug at two in the morning. After the town hall, Lena stayed behind and said, "Thank you. Now I understand exactly how little say I have." She was not being sarcastic. She meant it as the most precise description she could offer of what the transparency had actually done. Before, the system had been a black box, and a black box at least leaves room to imagine that your judgment might matter, that there is some human behind it who could be reasoned with. Now the box was glass, and through the glass Lena could see, with perfect clarity, that the thing scoring her work could not hear her and was never going to. The understanding had not given her a voice. It had shown her, in high resolution, that she did not have one.

That is the surprise at the heart of the fourth tension, and it undoes more well-meaning leaders than any other finding in this book. Opening the black box is necessary. It is also nowhere near sufficient. And done by itself, without the thing that has to come with it, it can make the very problem it was meant to solve measurably worse.

Priya did not give up on transparency, and neither should you, which is exactly what makes this tension so treacherous. The instinct that sent her to open the box was correct. A system nobody can see is a system nobody can trust, and she was right to refuse that. The mistake was believing that the opening was the whole job, that once the box was glass the work was finished. She had completed the first half of a two-part move and stopped, and stopping halfway through this particular move does not leave you halfway to trust. It can leave you further from it than when you began, which is the part that makes no sense at all until you see the mechanism underneath it.

## The Reasonable Theory That Fails

The standard prescription here is sensible, widely believed, and exactly half right, which makes it more dangerous than a prescription that is simply wrong, because the half that works lends credibility to the half that backfires.

People distrust what they cannot see, the theory goes, so show them. Reveal the logic, expose the weights, make the algorithm legible. Regulators reach for it, vendors advertise it, and good leaders implement it in good faith. And it does accomplish something real. A system you understand is one you can at least audit, contest in principle, and stop fearing as a malevolent mystery. Comprehension is a genuine good, and I would never argue for keeping people in the dark. The error is narrower and more specific. It is the belief that comprehension is the same thing as trust, or that it reliably produces trust on its own. It does not, and the gap between the two is where careful leaders fall.

The evidence on this is blunt. Across the reviewed studies, explainable systems raised trust meaningfully, but only when the explanation arrived together with voice. Where people received explanation and nothing else, no ability to respond, comprehension rose and trust did not. The explanation by itself was inert, and sometimes worse than inert. It is one of the cleaner findings in the whole literature, and it contradicts the instinct of nearly every leader who has ever tried to fix distrust by being more open, which is to say nearly every good leader there is.

## The Glass Box Problem

There is a strange asymmetry between a black box and a glass one, and it is the key to the whole puzzle.

A black box hides its workings, and people fear what they cannot see. But a black box also leaves room. Inside the not-knowing, a person can hold open the possibility that the system is fairer than it looks, that a human reviews its calls, that their own situation is understood by someone, somewhere. The ambiguity is uncomfortable, and it is also a kind of shelter for hope. When you make the box glass, you remove the discomfort and the shelter in the same motion. Now the person sees exactly what the system does, and if what it does is decide without them, the glass does not reassure. It confirms. The fear of an unfair process is replaced by the certainty of an indifferent one, and certainty of indifference is harder to live with than fear of unfairness, because fear can still be argued with and certainty cannot.

This is why Priya's transparency lowered trust in some teams instead of raising it. For the engineers who had quietly believed there was some give in the system, the glass took the give away. They had not been trusting the system. They had been trusting their own hopeful story about it, and the transparency, by replacing the story with the facts, dissolved the very thing that had been holding their trust up. You cannot build trust by destroying the illusions people were using in its place, unless you put something real where the illusion stood. The something real is voice, and Priya had not yet supplied it.

## Explanation Answers the Wrong Question

To see why, you have to notice that transparency and trust are answers to two different questions, and the person on the other end is asking both at the same time.

Explanation answers a question of the head. Do I understand what is happening to me? When the answer becomes yes, a real anxiety lifts, the particular dread of being judged by a process you cannot see. Call what explanation provides Explanatory-Assurance, the relief of no longer facing a mystery. It is worth having, and Priya's engineers did get it. But underneath that question sits a second one, a question not of the head but of standing. Am I a participant in this, or am I merely its subject? That question is not touched by explanation at all. It is answered only by voice, by whether the person has any ability to shape, contest, or talk back to the thing that is judging them. Call what answers it Participatory-Validation, the confirmation that your judgment actually counts for something. The two needs are different, and satisfying the first does nothing for the second.

Here is the trap, stated as cleanly as I can. Disclosure satisfies the first question and, all on its own, can sharpen the second into a wound. Before the transparency, Lena could tell herself a story in which her judgment might still matter somewhere in the system. The transparency replaced the story with a fact: it does not. The assurance arrived, and riding in on the very same delivery came a precise and unwelcome answer to the question she cared about far more. You have shown me the machine, and now I can see there is no door in it for me. Informed powerlessness is more corrosive than the ignorant kind, because ignorance at least leaves room for hope, and clarity removes even that. You have not reassured the person. You have confirmed their worst suspicion in excellent detail.

## Three Kinds of Transparency

It helps to separate three things that all travel under the single word transparency, because organizations reliably provide the first two and skip the third, which is the only one that builds trust.

The first is transparency of data: showing people what the system records about them, the numbers, the inputs, the score. The second is transparency of logic: showing them how those inputs become an output, the weights and the rules and the reasoning. Priya had delivered both, in abundance, and both are genuinely worth having. The third kind is transparency of recourse: showing people what they can actually do about any of it. What happens when they disagree. Where the door is. Whether pushing on it accomplishes anything at all. This is the transparency that answers the question of standing rather than the question of the head, and it is the one almost no system provides, because the first two can be satisfied with a document and the third can only be satisfied with a mechanism that costs the organization something.

Here is the trap that catches good leaders. Transparency of data and logic, supplied without transparency of recourse, does not read as openness. It reads as a confession. You have told the person exactly how they are being judged and, by your silence on the third point, exactly that they cannot affect it. The more thorough your disclosure of the first two, the louder your silence on the third becomes, which is the real reason the most transparent systems, in the narrow sense, sometimes produce the least trust. They have explained the verdict in immaculate detail and left no avenue of appeal, and a detailed verdict with no appeal is very nearly the definition of a thing people do not trust.

## Grace's Counterpoint

The contrast that proves the point came from Grace, who had stumbled into the right answer from the opposite direction without quite knowing how she had done it.

In one of her hospitals, instead of explaining a finished system to the nurses, the team had brought the nurses in before it was finished, to help shape what it measured and how it reported. And they had built in something small and decisive. A nurse could annotate the system's output, flag a score she disagreed with, and add the context the system had missed, and the annotation was visible and actually used when the system was tuned. It was not a comment box that emptied into nowhere. It was a writable pane, a place where a nurse's judgment could be entered into the record and have an effect.

The numbers told the story. Where workers could annotate, flag, or add context to the system's output, acceptance of that system rose by about forty percent. Trust climbed alongside understanding rather than diverging from it. It was the same transparency Priya had offered her engineers, plus the one thing Priya had left out, a way to write back. The difference between Lena's glass box and the nurses' system was not how much each group understood. Both understood perfectly well. The difference was that the nurses could talk back and be heard, and Lena could only watch. One group had been made participants. The other had been made spectators with very good seats.

Two things made the difference, and they are worth separating. The first was that the nurses had been there before the system was finished, which gave them a sense of ownership that no after-the-fact explanation can manufacture, because you do not feel like the subject of a thing you helped build. The second, and the more durable, was the writable pane, because ownership at the start fades unless it is renewed, and the pane renewed it every day. Every time a nurse flagged a score and saw the flag matter, the system reminded her that she was a participant and not a spectator. Co-design opened the door. The writable pane kept it from swinging shut, which is the part most organizations forget, because they treat involvement as an event rather than as a standing condition.

## The Threshold

The proposition for this tension carries one more variable than the others, and it is a cultural one that has to be handled with some care.

Where the system's logic is disclosed, and people have writable panes, the ability to annotate and contest, and the setting is relatively low in power distance, trust rises by about half a standard deviation. Power distance is the degree to which people in a group accept and expect steep hierarchy, the extent to which those lower down assume they should defer to those above without question. It is a feature of cultures, national and organizational, and it varies enormously, and it matters here because voice is only ever as useful as people's willingness to actually use it.

In a low-power-distance setting, hand people a writable pane and they will write in it. They will contest, argue, and improve the system, because they assume their input is welcome and consequential. In a high-power-distance setting, the identical writable pane may sit untouched, because people do not believe it is truly safe or appropriate to contradict the system, or the authority behind it, even when they have been explicitly invited to. So the lever is not add a comment box and you are finished. It is build the voice, and then build the conditions under which people will actually dare to use it, which in a steep hierarchy means leaders visibly inviting dissent, acting on it where they can, and doing it again and again until people come to believe the invitation is real. The full equation is transparency times voice times the freedom to use the voice, and the third term is the one leaders most often forget to supply.

Priya saw this directly when her company rolled the writable panes out across offices on three continents. In the offices where engineers were accustomed to challenging senior people as a matter of course, the panes filled with sharp, useful contest within days. In the offices where the culture ran more deferential, the identical panes sat nearly empty for weeks, not because those engineers had fewer disagreements but because they did not yet believe that writing one down was safe or welcome. The tool was the same everywhere. The willingness was not, and the willingness, not the tool, was what produced the trust. She learned to spend more leadership effort, not less, in exactly the places where the panes stayed quiet, because silence in a writable pane is not the absence of disagreement. It is disagreement that has decided to stay invisible, which is the most expensive kind there is.

## Why Trust Is the Multiplier

It would be a mistake to file trust, like autonomy and safety before it, under soft, because trust is the variable that decides whether any of the others ever pay off.

A system that people understand but do not trust does not get quietly ignored. It gets gamed. People route around it, feed it what it wants while starving the goal it was meant to serve, and optimize the metric against its own purpose, exactly as a sales team will in the very next chapter. A distrusted system turns its own users into adversaries, and an adversarial user is a deeply creative one, because no one knows the seams of a system better than the person it measures all day. You have handed your most knowledgeable people both a reason to defeat the system and a perfect map of how to do it.

A trusted system does the opposite. It turns its users into co-maintainers. They flag its errors instead of exploiting them, feed it better information than it could ever gather alone, and improve it steadily over time. Trust is the thing that decides whether the people who know the most about the work, the ones standing on the floor, spend that knowledge improving your system or quietly defeating it. Which means transparency without voice is not merely insufficient. It is actively expensive, because it produces people who understand the system well enough to game it expertly and have been given no reason at all not to.

## The Gaming Spiral

Watch what happens to a system that is understood and distrusted, left to run, and you see why trust is not a nicety but the thing that determines whether the system survives contact with its own users.

It begins with small gaming. People learn the seams, which the transparency has helpfully shown them, and they optimize the metric in ways that serve the score and not the goal. Management notices the gaming and responds the only way an untrusting management knows how, by adding rules to close the loopholes. The new rules create new seams, which the users find and exploit, because nothing has changed about their reason to. Management adds more rules. The system grows more complex, more rigid, and more adversarial with every turn, an arms race between the people who set the metric and the people who live under it, and the work itself, the actual goal everyone supposedly shares, is the casualty that no one is defending. Each side is behaving rationally. The spiral is the product of the distrust, not of bad people, and it does not stop until something breaks the cycle, which is almost always a restoration of voice.

The contrast with a trusted system is total. In a trusted system, the same engineer who would have gamed the metric instead flags the loophole, because he believes that flagging it will lead somewhere better than exploiting it. The user's knowledge of the seams becomes an asset instead of a weapon. That is the entire return on trust, and it is why transparency without voice is such a bad trade. It hands your most capable people a perfect map to every seam in the system, and gives them no reason at all to use it for you rather than against you.

Marcus and Grace had both learned this from the other side. Marcus's drivers, before calibration, had trusted his routing system so little that they kept private workarounds, little deviations they never logged, which meant his data was quietly fiction. Grace's nurses, before co-design, had trusted the early-warning scores so little that they double-checked everything by hand, which erased the very efficiency the system was supposed to deliver. In both cases the distrust did not announce itself as distrust. It showed up as bad data and lost time, costs that no one traced back to the missing voice that had caused them. Trust is the variable that decides whether the people closest to the work are feeding your system or fooling it, and a fooled system is worse than no system at all, because it is confidently wrong and expensive to maintain in its wrongness.

## What Priya Tried Next

Priya's real fix was the thing she had left out the first time, and it turned out to be smaller than she had feared. She gave them a way to write back.

She added writable panes to the scoring system. An engineer who disagreed with a score could annotate it, attach context, and contest the call, and, crucially, those contests did not drop into a void. They were reviewed by a human, some of them changed the score, and the patterns across them fed the ongoing tuning of the system itself. Trust rose, and it rose not because the engineers won every contest, because they certainly did not, but because the contest existed and was real. Lena did not need to be right every time. She needed the door to exist, and to swing when she pushed on it hard enough and well enough.

The reframe Priya carried into everything afterward was this. Transparency is the floor, not the building. It is necessary, and you stand on it, and a building with no floor is no building at all. But trust is constructed above the floor, out of voice, and the structure simply does not exist until you build that part. The two levers that carry this furthest, explainable dashboards done properly and the moderated forums where voice becomes systematic rather than occasional, are waiting in Part Three. For now the principle is enough. Show people the machine, yes, but then give them a door into it, because a person who can see in and cannot get in does not feel trusted. They feel watched through glass, and no one in the history of the world has ever trusted the thing that watches them through glass.

One contest in particular changed how Priya thought about the whole thing. An engineer flagged that the system was penalizing a pattern of code that was, in one specific subsystem, exactly the right thing to do. The flag was correct, the system was adjusted, and the story traveled. After that, engineers contested more rather than less, because the first successful contest had proven the door was real, and a door that has been seen to open is one people will keep trying. But Priya also learned to bound the voice, because a writable pane is not an infinite veto. Engineers could contest and be heard. They could not simply overrule the system by objecting to it. Some contests changed the score and many did not, and the legitimacy of the whole arrangement depended on the contests being judged honestly rather than always granted. Voice is not the same thing as getting your way. It is the right to be genuinely heard by someone with the authority to act, and people can tell the difference between that and a suggestion box, almost immediately.

The principle to carry out of Priya's second tension is short and easily forgotten. Disclosure earns you understanding. Only voice earns you trust, and the two are not the same currency, no matter how much the first feels like it should buy the second.

Priya had her door now, and her engineers had begun, slowly, to walk through it, contesting and annotating and, in the doing, becoming the co-maintainers a distrusted system never gets. The glass box had become a workshop, which is what the transparency was supposed to produce all along.

Those are the four tensions. Specificity against autonomy, cadence against safety, efficiency against learning, transparency against trust. Four faces of one paradox, each diagnosed, each with its mechanism named and its threshold marked. The diagnosis is finished. But a diagnosis that does not lead to a treatment is only a more sophisticated way of worrying, and Marcus, Grace, and Priya had not called me in order to worry more precisely. They had called me to fix it. So we turn now from naming the problem to building the answer. There are five levers, and the first one belongs to Marcus, who had spent a year learning that you cannot fix too much precision with less precision, and had finally found what you fix it with instead.

### THE EVIDENCE

Transparency against trust appeared in forty-two percent of the reviewed studies. Explanation alone, the Explanatory-Assurance face, raised comprehension but left trust flat and sometimes lowered it. What moved trust was Participatory-Validation: where workers could annotate or contest the system's output, acceptance rose by about forty percent.

The proposition: where logic is disclosed, people have writable panes, and power distance is relatively low, trust rises about 0.50 SD. The third condition matters. In steep hierarchies a writable pane sits unused until leaders prove, repeatedly, that contesting the system is safe and consequential.

### MONDAY MORNING

Find a system your people understand and still distrust, then ask one question. Can they talk back to it, and does anything actually happen when they do? If the answer is no, you do not have a transparency problem. You have a voice problem, and more explanation will make it worse rather than better. Add one writable pane, on one system, route the responses to a human who can act, and watch what your people do with it.

PART THREE

# The Five Levers

CHAPTER EIGHT

# Goal Calibration

*A goal handed down is a fence. A goal agreed to is a promise. People climb fences and keep promises.*

A year after Carmen, Marcus stood in a room full of drivers who were arguing with him, and it was the best meeting he had run in a decade.

The six-driver experiment from the previous year had grown into something the whole network now ran. Once a month, on every site, a mixed group of drivers and warehouse leads, veterans beside newcomers, sat down with the routing and scoring rules projected on a screen, not as a finished system to be briefed on but as a draft to be argued with. On this particular morning the argument was about a prompt that pushed drivers to skip a particular consolidation step in order to shave minutes. Two veterans were insisting the step was the only thing keeping a certain class of fragile shipments from arriving broken, and the dashboard had no way to see breakage that happened three days later at the customer's dock. They were right. By the end of the hour the rule had changed. Marcus walked out understanding something about his own operation that no analyst had surfaced in two years, and he had learned it from the people the system spent all day instructing.

What struck Marcus most was not that the rule changed. It was the room. A year earlier these same drivers would have sat in silence through any meeting he called, answering direct questions in as few words as possible and waiting to be released. Now they argued with him, and with each other, and the veterans pulled the newcomers into it, and nobody was performing for the dashboard, because the dashboard was the thing they were there to fix rather than the thing watching them. He had spent a career trying to get frontline people to bring him the truth about his own operation, and he had finally found the trick, which was to stop asking them to approve his decisions and start asking them to improve them.

For six chapters we have been naming the disease. The four tensions are the diagnosis, and a careful one, but a diagnosis is not a cure, and no one was ever made well by being told more precisely what was wrong with them. Now the treatment. The research points to five levers, and the most important thing to understand about them before we begin is that they are not five separate fixes you choose among. They are sequenced. Each one builds the capacity the next one needs, and run together they form a loop an organization keeps turning rather than a project it finishes and files away. The first lever is the one that answers Marcus's tension, and it is the one most leaders reach for last, because it feels like the one most likely to cost them control. It is, in fact, the opposite. It is goal calibration.

## What Goal Calibration Is

Goal calibration is the practice of bringing the people who do the work into the setting, and the ongoing adjustment, of the targets the system optimizes.

Notice what it is not, because every word of the definition is doing work. It is not abandoning metrics, and it is not lowering the bar, and it is emphatically not a vote. It is co-authorship of the targets, on a regular cadence, with the leader still holding the pen the entire time. The system keeps measuring exactly as before. The question calibration asks is whether the things it measures are the right things, weighted the right way, aimed at the actual goal rather than a convenient proxy for it. And the people best positioned to answer that question are not the analysts who built the dashboard. They are the people whose work is being measured, who can see every single day what the metric is missing, rewarding by mistake, or quietly punishing. Calibration is the channel that connects what they can see to what the system counts.

## The Fear of Losing Control

The reason most leaders reach for this lever last, if they reach for it at all, is a fear worth taking seriously and then dismantling. They are afraid that bringing workers into the targets means losing control of them.

It feels like control to set the targets yourself, hand them down, and hold people to them. It feels like a loss of control to open them up. But this gets the accounting exactly backward, and the previous chapters explain why. A target that people do not own is a target they will game, route around, or quietly resent into ineffectiveness, and a gamed metric is not control. It is the appearance of control sitting on top of an operation that is doing something other than what you believe you measured. The bell rang on every closed deal and management felt fully in control right up until they discovered the team had stopped selling the deals that mattered. That is what losing control actually looks like, and imposing the metric is what produced it.

Real control is not the power to set a number. It is the confidence that the number means what you think it means, and that people are actually pursuing the goal underneath it. Calibration increases that kind of control, because it aligns the metric with the goal and enlists the people who can see the gap between the two. You give up a little of the feeling of control, the unilateral authorship, and you gain a great deal of the substance of it. Leaders who cannot tell the feeling from the substance will keep choosing the feeling, and they will keep being surprised by what their well-controlled operations turn out to have been doing all along.

## Why It Works

It works on two levels at once, which is precisely why it is the right lever to start with.

First, it restores the autonomy that the precision eroded, without giving up the precision. Recall the core of the first tension. The very act of defining the target compresses the worker's discretion, because the more exactly the system says what to do, the less the worker decides. Calibration returns discretion at a higher level. The driver may not choose each turn, but he has a hand in choosing what the turns are measured against, and that restores the sense of agency the basic human need for autonomy actually requires. A driver who helped decide which prompts matter is exercising judgment about the system even as he follows it. The order has quietly become an agreement, and a person can stand inside an agreement in a way they never can inside an order.

Second, and this is the part the hardheaded operator should care about even if the first leaves them cold, it makes the metrics measurably better. The people on the floor know what the dashboard cannot see. They know the Carmen column, the work that matters and goes uncounted. They know the perverse incentive the metric is quietly creating, the consolidation step being skipped, the fragile shipment being broken three days downstream where no sensor is watching. Calibration is the channel through which that knowledge reaches the metric and corrects it. You are not merely making people feel heard, though you are doing that too. You are fixing measurement error using the only instrument capable of detecting it, which is the human being standing in the gap between the number and the work it is supposed to represent.

There is a third effect, quieter than the other two and slower to arrive, and it is the reason calibration comes first among the levers rather than somewhere in the middle. It builds trust, in the specific and consequential sense the previous chapter defined. Every time a worker speaks into the calibration and sees the metric change, or hears an honest reason it cannot, the organization proves that voice is real here. That proof is the foundation the other four levers are built on. Hybrid coaching, rotational upskilling, explainable dashboards, voice forums, every one of them asks people to engage honestly with a system that holds power over them, and people will only do that inside an organization that has shown them honesty gets answered. Calibration is where the showing starts, which is why a leader who skips it and reaches for the later levers usually finds them strangely inert. They have tried to build the upper floors before pouring the foundation.

## How to Run It

The practice is simple to describe and takes real discipline to sustain. Here is the shape Marcus settled into, and it generalizes to almost any operation.

Step one, pick the contestable metrics. Not everything is open for calibration, and pretending otherwise is its own kind of dishonesty. Some targets are fixed by safety, by law, or by a strategy that is genuinely not up for discussion, and you say so plainly and without apology. But most operational metrics carry buried choices, the weights, the thresholds, what counts as a success and what does not, and those choices are exactly where calibration lives. Put the contestable ones on the table, and be just as clear about which ones are not, because a calibration that pretends everything is negotiable collapses the moment it has to enforce something.

Step two, bring the right people, and show them the metric as a draft rather than a verdict. The people in the room should be the ones the metric actually measures, a deliberate mix of veterans and newcomers, and the metric should be presented as something they are being asked to improve, not something they are being asked to accept. The framing is very nearly everything. Help me get this right produces argument, which is what you want. Here is how this works produces silence, which is what you are trying to escape. The same hour, the same people, the same data, and the only variable is whether you walked in asking a question or delivering an announcement.

Step three, ask three questions, and then be quiet long enough to hear the answers. Which of these targets actually helps you do the work well? Which one insults a judgment you have already made correctly, or pushes you toward something you know is wrong? And what is the system measuring that it should not, or failing to measure that it should? These three questions, asked of the people who know, will surface more real measurement error in a single hour than a quarter of analyst review, because the analysts are reasoning about the work from a distance and the people in the room are living inside it.

Step four, change what you can and explain what you cannot, in the room, on the same day. The fastest way to kill calibration stone dead is to gather input and let it vanish into a process no one ever hears from again. Some suggestions you will be able to adopt on the spot, so do it visibly, while they watch. Some you will not be able to act on, for reasons the room may not be able to see, so say why, specifically and honestly. The entire credibility of the practice rests on this single step. People will forgive a no that they understand. They will never forgive a black hole, and one black hole will teach them, permanently, that the whole exercise was theater.

Step five, make it recurring, and open every session by closing the last one. Calibration is not a launch event you hold once and congratulate yourself for. It is a standing cadence, monthly or quarterly, and every session begins by reporting what actually changed because of the session before it. That loop closing is what converts a one-time consultation into a durable belief that voice is real and consequential, which is, not coincidentally, exactly the belief the previous chapter identified as the foundation of trust. Calibration done once is a survey. Calibration done forever is a relationship.

## Who Should Be in the Room

The practice lives or dies on who is in the room, and the obvious choices are not always the right ones.

Bring the people the metric measures, which sounds simple and is routinely gotten wrong, because the people who volunteer and the people who speak loudest are not a representative sample of what your floor actually knows. Marcus learned to seek out the quiet veterans, the ones who never raised their hands and who held the deepest models of how the work really went, and to protect them in the room from the few who would otherwise dominate it. He learned to mix tenure deliberately, because the veterans knew what the metric missed and the newcomers knew what the metric felt like to someone still forming their judgment, and the two kinds of knowledge corrected each other. And he learned to keep the group small enough that silence was conspicuous, because in a room of forty no one speaks, and in a room of eight everyone eventually must.

Above all he learned that the room had to be safe in exactly the sense Grace's wards had not been. If contesting a metric in the calibration session could come back to bite a worker in their own score, the session would fill with the same careful silence that had nearly cost Grace a patient. So the calibration room was explicitly walled off from evaluation. What you said about the metric could change the metric. It could never change your rating. Without that wall, calibration is not a channel for truth. It is just one more stage, with one more audience, and people perform on it exactly as they perform everywhere else they are being scored.

## The Bell That Backfired

A short story from outside Marcus's world, because the failure mode it shows is so common that it deserves a name and a warning.

A software sales team I once worked with had installed a dashboard that rang a bell, an actual sound on every screen in the office, each time a rep closed a deal. Management loved it. It was visible, it was motivating, it gamified the floor, and for a few weeks it genuinely lifted energy. And within two quarters it had quietly wrecked the team, in a way that took everyone by surprise because the number of deals closed, the thing the bell celebrated, had gone up the whole time.

The reps did what people always do with a metric they did not help set and do not trust. They optimized it against its own purpose. The bell rewarded closed deals, so the reps cherry-picked the small, fast, easy ones and quietly let the large, slow, strategic ones wither, because a major account that took two quarters to land rang the bell exactly once, the same as a trivial deal that closed in a week. The best reps, the ones who patiently worked the hard and valuable deals, started looking worse on the board than the ones racking up a noise of tiny wins. One of them, a quietly excellent rep named Jo, nearly quit over it. If that sounds familiar, it should. It was Carmen again, wearing a sales lanyard. The metric could see the bell and could not see the business.

Calibration fixed it, and the fix came from the reps rather than from management. Brought into the room and asked the three questions, they said in about five minutes what the analysts had missed for two quarters. The bell measures the wrong thing. It measures deals closed, not value built, and it is training us to abandon precisely the customers you most want us to win. Together they redesigned it, weighting for deal size and strategic fit, retiring the bell in favor of a quieter signal that did not reward speed over worth. The metric had been wrong from the start, the people who could see it were the very people it measured, and calibration was the only channel that could carry what they knew to the place where the number was set.

Jo stayed, which is the part of the story I care about most. The rep who had nearly quit over a metric that punished her for doing the valuable work became one of the strongest advocates for the recalibrated system, because she had helped build it and it now measured something she could be proud of hitting. That is the quiet dividend of calibration that never shows up on a dashboard. The metric you build with your best people is a metric your best people will defend, and the ones you were most at risk of losing become the ones most invested in the thing you fixed. You do not only repair the measurement. You re-recruit the very people the broken measurement had been driving away.

## What Calibration Is Not

Because the lever is so easily misunderstood, and so easily turned into something soft and useless, it is worth being firm about its edges.

Calibration is not consensus, and it is not democracy. The leader holds the pen from the first minute to the last, and everyone in the room should know it. It is not lowering the bar either. Run well, it usually raises the bar, because people defend targets they helped set far more fiercely than targets imposed on them, and a goal you chose is a goal you will fight to hit. It is not a morale program, and this is the most important boundary of all. If you run calibration to make people feel good rather than to make the metrics true, they will see through it in a single session, and the disillusionment it produces will do more damage than the silence it replaced. And it is not the abdication of judgment to the crowd. It is the disciplined gathering of the one kind of information the system cannot collect on its own, filtered through a leader who still decides. Authorship, not abdication. The pen does not leave your hand. You simply stop pretending you are the only person in the building who can see the page.

Held to those boundaries, calibration becomes the rarest thing in a measured organization: a place where telling the truth about the numbers is rewarded rather than punished. That alone, in most companies, would be worth the trouble even if it did nothing else.

## When Calibration Fails

Calibration fails in four recognizable ways, and a leader who knows the four can usually steer around them.

It fails as theater, when input is gathered and nothing visibly changes, which teaches people faster than any speech that the exercise is a performance of listening rather than the real thing. It fails by capture, when the loudest or most senior voices in the room dominate and the metric ends up serving them rather than the work, which is why who is in the room and how it is run matter so much. It fails by scope creep, when everything is treated as negotiable and the session dissolves into a complaint forum with no pen behind it, which is why naming the non-negotiables up front is not optional. And it fails as a one-time event, when a leader runs a single celebrated session, banks the goodwill, and never returns, which converts a relationship back into a survey and spends the very trust it briefly built.

Notice that three of the four failures are failures of follow-through rather than of design. Calibration is not hard to start. Almost any leader can run one good session and feel the energy in the room. It is hard to sustain, because sustaining it means changing real metrics on a real cadence and closing the loop every single time, forever, which is a standing tax on a leader's attention. That tax is the actual price of the lever, and leaders who are unwilling to pay it should not start, because a calibration that is begun and abandoned does more damage than one never attempted. It teaches people that their voice was wanted once, for show, and is not wanted now.

## The Evidence Read Forward

The evidence behind calibration is the evidence behind the first tension, simply read forward into action.

Recall the first proposition. The autonomy loss from precise, fast-refreshing systems was conditional, and the conditions that buffered it included voice and readiness. Calibration is, precisely, the deliberate manufacture of voice over the targets, and it raises the readiness that buffers the harm, because it teaches people to engage the system as something they shape rather than something that merely happens to them. Where calibration is real and recurring, the autonomy cost of precision falls toward zero while the precision's gains remain fully intact. That is the dynamic equilibrium of the first tension, the holding of both poles at once, made operational at last. You keep the twelve percent accuracy. You stop paying for it in discretion. And you do it not by weakening the system but by giving its subjects a hand on the dial.

The principle to carry out of the first lever is the epigraph turned into an instruction. A goal handed down is a fence, and a goal agreed to is a promise. If you want people to own the numbers, give them a real hand in choosing the numbers, and keep the pen in yours. The two acts are not in tension. They are the whole of the craft.

## Calibration and the Machine

There is a frontier worth naming before we leave the first lever, because it connects calibration to a lever still ahead.

So far calibration has been a human practice, a room and a cadence and a set of three questions. But the writable panes from the last chapter point at something more. The contests people enter, the flags they raise, the adjustments the room agrees to, are all data, and increasingly they can be fed back into the system itself, so that the algorithm learns not only from the work but from the workers' judgments about how the work should be measured. Done carelessly this is just an automated suggestion box, and it fails the same way every suggestion box fails. Done well, with the human room still holding the pen, it turns calibration from a periodic correction into a continuous one, the metric and the people who live under it adjusting each other in something close to real time. That is the direction the explainable-dashboard lever will eventually take us, and it is why these levers form a loop and not a list. The voice you build here becomes the signal the system learns from later.

Calibration restores the worker's voice over what is measured. But voice over the target does not, by itself, help a person in the charged moment when the system's feedback lands and they have to decide what to do with it. That moment, the instant of feedback, is where the second lever lives, and it belongs to Grace, who learned that the most powerful thing you can add to an algorithm is not more algorithm. It is a human being, at exactly the right moment, asking one good question.

## THE EVIDENCE

Goal calibration is the first of the five levers and the primary answer to specificity against autonomy. It returns discretion at the level of what is measured, satisfying the autonomy need without surrendering the precision that drives the performance gain.

It draws directly on the first proposition, where voice and readiness buffered the autonomy loss. Calibration is the deliberate construction of voice over the targets, and closing its loop builds the consequential-voice that the fourth tension showed is the true foundation of trust. The levers are sequenced; this one builds the trust the rest will need.

## MONDAY MORNING

Pick one metric your people quietly resent. Put it in front of them as a draft and ask the three questions: what helps, what insults your judgment, what is it missing. Commit, before you walk into the room, to changing at least one thing they tell you and to explaining why on anything you cannot. You will learn more about your own measurement error in one hour than your last quarter of dashboards taught you, and your people will learn that the pen, while still in your hand, is finally writing some of what they say.

CHAPTER NINE

# Hybrid Coaching

*A number tells you where you stand. It takes a person to tell you what to do about it, and whether to believe it at all.*

The flag went red on Bisi's screen at the worst possible moment, in the middle of a shift that was already coming apart, and six months earlier she would have done one of two wrong things with it.

Bisi had been a nurse for two years. The patient in question, an older man recovering from pneumonia, had just been scored a rising risk by the early-warning system. On a normal day Bisi might have dismissed it, because she had learned, like everyone on the floor, that the system flagged more than it needed to, and a nurse who chased every flag would never finish a shift. Or she might have escalated in a small panic and called for help that was not warranted, which carried its own real costs. Either way she would have been reacting to the flag rather than thinking about it. The flag would have been in charge of her, and she would have been its hands.

Instead she found Ada. This was the new thing on Grace's floors, and it had grown out of the protected pauses Grace had started after the near miss with bed nine. The most experienced nurses had been given a second role. They were signal coaches, and their job was not to tell a newer nurse the right answer but to stand beside her while she worked it out. “Okay,” Ada said, looking at the screen with her, “the system flagged him. What is it seeing?”

Bisi looked. A rising heart rate, a small drop in oxygen, a temperature creeping up. “What might it be missing?” Ada asked. Bisi thought. The man had just been moved from his bed to a chair for the first time in three days, which would lift his heart rate and his temperature on its own, and he had pulled his oxygen line slightly askew, which she fixed as she stood there, and the number corrected in front of them. “So what do you think?” Ada asked. Bisi said she thought it was the movement and the line, not a true deterioration, but that she would watch him closely for the next hour to be sure, and document why. That was the right answer. More than that, she had reached it herself, with the system’s signal as one input among several rather than as a command she had to obey or defy.

Here is what had just happened, and it is the whole of the second lever. A year before, the same flag would have produced either blind obedience or blind dismissal, and neither of those is judgment. What Ada added was not a better algorithm, or a louder one, or a quieter one. It was a step the system had silently removed, the step where a human being interprets the signal before acting on it. Grace had discovered that the most powerful thing she could add to her algorithm was not more algorithm. It was a person, at the right moment, asking one good question.

Bisi did not know it, but she had just done something the system was incapable of doing and the old way of managing used to do without thinking. She had held the machine's judgment and her own in the same hand at the same time, given each its proper weight, and come out with a decision that belonged to neither alone. That is what judgment is, and it is the thing that quietly disappears when a signal runs straight into a reaction with no one in between. Ada had not taught Bisi the answer. She had taught her the motion, the holding of two judgments at once, which is a thing a person can only learn by being walked through it until the motion becomes their own.

## The Missing Step

Every algorithmic feedback system runs the same loop. The system observes, the system signals, the worker reacts.

Look closely at that loop and you notice a step has gone missing, the step that used to sit between the signal and the reaction, where a person interpreted what the signal meant in context before deciding what to do about it. When the loop ran through an experienced human manager, that step happened automatically and invisibly, because interpreting signals is most of what an experienced manager does. When the system became the manager, the step was removed, because the system signals and the worker reacts and there is no interpreter standing in between. And a worker with no interpretation step has only two moves available. Obey the signal, or ignore it. Neither is judgment, and judgment is the thing you actually needed.

This is why the cadence tension does what it does. A signal arriving every two minutes, with no interpretive step, trains people into one of the two reflexes. The anxious comply, chasing every flag, performing for the screen, going quiet like Ada in the first chapter she appeared in. The experienced learn to dismiss, tuning the system out as noise, which is its own kind of danger, because the one flag in fifty that mattered is now lost in the forty-nine they have trained themselves to ignore. Both responses are failures of judgment, and both are exactly what you get when you take a feedback loop that was designed for a human interpreter and run it straight from a machine into a human reactor with the interpreter deleted. Hybrid coaching is the deliberate reinstallation of the missing step.

## What Hybrid Coaching Is

Hybrid coaching is the pairing of algorithmic feedback with periodic human coaching whose purpose is to help the worker interpret the signal, weigh it against their own judgment, and decide.

The word hybrid is doing the work, so it is worth dwelling on. This is not human coaching instead of the algorithm, which would throw away the speed and consistency the system genuinely provides and which no operator under real pressure will accept. And it is not the algorithm instead of human coaching, which is precisely the arrangement that produced every problem in this book. It is the two together, with the algorithm supplying the signal and the human supplying the interpretation. The machine is fast and tireless and blind to context. The coach is slow and scarce and full of context. Each is strong exactly where the other is weak, which is the entire reason to run them as a pair rather than as rivals.

And the coach does not deliver verdicts. This is the part that separates real hybrid coaching from a manager simply explaining the dashboard. The coach asks the questions that rebuild the worker's own ability to read the signal, the what-is-it-seeing and what-might-it-be-missing and what-do-you-think, until the worker can eventually do the interpreting alone. The goal is not a worker who depends on the coach. It is a worker who has internalized the coach, who carries the interpretive step in their own head and no longer needs anyone standing beside them to perform it. A coach who creates dependence has failed. A coach who makes themselves unnecessary has done the job exactly right.

## Three Modes of Coaching

Coaching against an algorithm happens in three modes, at three different distances from the moment, and a complete practice uses all three.

The first is coaching in the moment, beside the worker as a live signal arrives, which is what Ada did with Bisi. It is the most powerful and the least scalable, and it should be reserved for the high-stakes flags where a real decision is being made in real time. The second is coaching in review, sitting down afterward to look back at a set of signals and the choices made about them, asking what the worker saw and what they might have missed, with the pressure of the live moment removed. Review coaching scales far better, because it does not have to happen at the instant of the event, and it is where most of the deliberate skill-building actually occurs.

The third mode is coaching in pattern, stepping back periodically to look not at a single signal but at the whole shape of a worker's interaction with the system over weeks. Is this nurse chasing every flag, the anxious-compliance failure? Is she dismissing too many, the experienced-dismissal failure? The pattern reveals which of the two reflexes a person is sliding toward, before it hardens into a habit, and it is invisible at the level of any single signal. A coaching practice that uses only the first mode burns out its veterans and reaches almost no one. One that uses all three spends the scarce in-the-moment attention where it truly counts and does the bulk of its work in review and in pattern, where it scales.

## Why It Works

It works for three reasons, and they map precisely onto three of the harms we have catalogued, which is not a coincidence. The lever was reverse-engineered from the damage.

First, it restores psychological safety, because the coach is explicitly not the evaluator. The coaching conversation is off the record, on the worker's side, a place where it is safe to be uncertain out loud, which is the exact thing the spotlight had made impossible. When Bisi tells Ada she is not sure what the flag means, nothing happens to her score, because Ada has no part in her score. By separating the person who helps you from the system that judges you, hybrid coaching hands back the safety the cadence took away, and it does it without slowing the cadence at all. The fast signal keeps coming. It just stops being the only voice in the room.

Second, it rebuilds the metacognition the system removed. Recall that the erosion works in part by taking over the worker's self-assessment, the judging of one's own work, which is where a particular and irreplaceable kind of competence is built. The coach does not give that judgment back by handing over answers, because handed-over answers build nothing. By asking what is it seeing, what might it be missing, what do you think, the coach makes the worker perform the self-assessment the system had been performing for her, and the muscle regrows through use, the way any muscle does. Coaching in this sense is the deliberate re-creation of the very practice the efficiency lever crowds out, aimed precisely at the moment of feedback where the practice matters most.

Third, it catches the system's errors. A worker reacting alone to a flag cannot easily tell a true signal from a false one, and so she either trusts all of them or none of them. A worker and a coach together, reading the signal against what they both know of the situation, catch the misreads, the Carmens and the Marisols the system was about to punish for excellence it could not see. The coaching layer is, among everything else it does, an error-correction layer for the algorithm, and it is staffed by the only sensors in the building capable of seeing what the algorithm cannot. You are not just developing the worker. You are debugging the system, using the one instrument that can detect the bug.

There is a fourth effect, quieter than the others, and it is about the coach as much as the coached. Asking the interpretive questions keeps the veteran's own judgment sharp, because teaching a skill is one of the most reliable ways to deepen it, and because the questions a newer nurse cannot yet answer often expose assumptions the veteran had stopped examining years ago. Coaching is not a one-way pour from the full cup into the empty one. It is a practice that develops both people, which matters enormously for the scaling problem, because it means the time a veteran spends coaching is not subtracted from their own development. It is added to it.

## The Marisol Case

The clearest example I know came from a customer-service floor, a long way from any hospital, which is part of why it is worth telling.

Marisol was a veteran agent whose numbers were bad. The system scored agents on call duration and script adherence, and Marisol's calls ran long and wandered off-script with some regularity. By the dashboard she was one of the weakest agents on the floor, and she was a few weeks from a performance plan that would very likely have ended with her gone. Everyone who actually worked near her knew she was extraordinary. The system, which did not work near her, knew only that her calls were long and her adherence was low.

A coaching layer, added almost as an experiment, looked at Marisol's numbers with her rather than at her, and the entire picture inverted inside a single conversation. Her calls ran long because she solved the customer's whole problem on the first call, including the problems they had not thought to ask about yet, which meant her customers did not call back. Her off-script wandering was her diagnosing root causes the script had never anticipated. She had the lowest repeat-call rate and the highest satisfaction scores on the entire floor, and the system, measuring duration and adherence, had been faithfully reading her excellence as inefficiency. It was Carmen again, and Jo again, the best person scored worst by a system that could not see the actual work. The coaching caught what the dashboard could not, and Marisol, instead of being managed out, was made a coach herself, teaching the floor the root-cause approach her own numbers had been hiding. The metric had been about to fire its best teacher.

## The Coach as Sensor Network

There is a second thing coaching produces, almost as a byproduct, and it turns out to be valuable enough to justify the practice on its own.

When coaches across a floor keep encountering the same misread, the same kind of flag that keeps turning out to be wrong in the same way, they have discovered something about the system that no individual worker reacting alone could ever see. One nurse who notices the system over-flags patients newly moved to a chair has an anecdote. Forty coaches who all notice it have a finding, and that finding is exactly the kind of systematic measurement error the calibration room exists to fix.

The coaching layer, in other words, is not only an interpreter for the worker. It is a distributed sensor network for the system's blind spots, and its observations are the richest possible input to goal calibration.

This is the first place the levers visibly begin to feed one another. Coaching catches the patterns of misread. Calibration uses those patterns to correct the metric. The corrected metric throws fewer false flags. The coaches are freed to focus on the genuinely hard cases rather than the systematic noise. A leader who runs coaching and calibration as two separate programs gets two useful things. A leader who wires the output of the first into the input of the second gets a system that learns, which is worth far more than the sum of the parts, and which is the shape the whole of Part Three is quietly bending toward.

## How to Build It

The practice has a few rules that determine whether it works or quietly curdles into something worse than nothing.

First, the coach must not be the evaluator. This is the rule everything else depends on. If the person coaching you is the person who scores you, the conversation collapses back into surveillance, the safety evaporates, and you have simply added a friendlier-looking interrogation to the worker's day. Use experienced peers rather than the management chain, and wall the coaching off from evaluation as completely as Marcus walled off his calibration room. The worker has to know, with certainty, that nothing they say to the coach can reach their rating, because the moment they doubt it, they will go back to performing, and a performance is the one thing coaching cannot use.

Second, the coaching is about interpretation, not answers. The temptation, especially for an expert coach, is simply to tell the worker what to do, which is faster, feels helpful, and rebuilds nothing at all. The discipline is to ask rather than tell, to make the worker do the reading even when watching them do it slowly is almost unbearable for someone who already knows the answer. The slowness is not a cost of the method. It is the method. You are not solving today's flag. You are growing a nurse who can read tomorrow's flag with no one beside her, and that nurse is built only by being made to do the interpreting herself, today, while it is still safe to get it wrong.

Third, it has to be regular and protected, a standing part of the work rather than something that happens when there is time, because there is never time, and a practice that depends on slack will not survive its first busy week. And it should focus where it matters most, on the high-stakes signals and the moments of genuine uncertainty, not on every flag, which would be both impossible and pointless. The art is in choosing the moments. A good coaching system spends its scarce human attention on the handful of signals where interpretation changes the outcome, and lets the routine ones pass without ceremony.

## What the Coach Is Actually Doing

Underneath the questions, the coach is performing a specific and well-understood teaching move, and naming it helps you do it on purpose rather than by luck.

The move is to take a mental process that has become invisible and make it visible, then hand it over by stages. An expert nurse interprets a flag without conscious effort, because the interpreting has sunk below awareness, which is what expertise feels like from the inside. The coach's job is to drag that buried process back up into the open, to externalize it as a sequence of questions, what is it seeing, what might it be missing, what do you think, so that the novice can watch the thinking happen and then practice it themselves. Educators call this the gradual release of responsibility. The coach does it first while the worker watches, then alongside the worker, then watches while the worker does it, then steps away. At each stage the scaffolding is reduced, until the worker is doing alone what they could once do only with help.

This is why telling, however efficient it feels, builds nothing. A told answer transfers the conclusion and hides the process, leaving the worker exactly as unable to reach the next conclusion as they were before. The asked question forces the worker to run the process themselves, which is the only thing that builds the capacity to run it again next time, alone. The discomfort an expert feels watching a novice work slowly through something the expert could have done in an instant is the precise sensation of teaching working. If it feels efficient, you are probably doing it wrong, because you have probably started telling.

## The Coach Problem

The obvious objection is that coaching does not scale. You cannot put a veteran beside every worker for every signal, and veterans are exactly the scarce resource you are trying to protect. It is a fair objection, and it has three answers.

The first is that you do not need one-on-one coaching for every moment. Most of the value comes from group huddles, peer coaching in pairs, and the occasional well-timed conversation at a high-stakes moment, not from constant individual attention. A single experienced nurse running a fifteen-minute huddle reaches a whole shift, and much of the interpreting that matters is contagious once a few people on a floor start doing it out loud. Coaching spreads as a habit, not only as a transaction, and the habit is cheaper than the transaction.

The second answer is that the tools themselves can help, which is the hybrid idea cutting in the other direction. The same system that flags patients can flag moments, telling you where a human conversation would pay off most, so that scarce coaching attention flows to where it matters rather than being spread evenly and thinly across everything. The algorithm is genuinely good at triage. Let it triage the coaching. A system that can say this is the flag where a newer nurse most needs a veteran beside her is a system that has made its own coaching layer affordable.

The third answer, and the one that closes the loop with the next chapter, is that coaching is self-replenishing when you build it right. The nurse Ada coaches becomes, in a few years, a nurse who coaches. Marisol, caught and developed, became a coach. The practice that appears to drain your veterans is in fact the practice that manufactures your next ones, provided you treat coaching itself as a skill that people rotate into and out of, rather than a permanent tax on your most experienced few. That is more than an efficiency. It is the bridge to the third lever, which is built entirely on the idea of rotating people deliberately through the skills you cannot afford to lose.

Put the three answers together and the objection dissolves. You target the scarce in-the-moment coaching at the high-stakes flags, you let the algorithm tell you where those moments are, and you treat coaching as a skill that propagates, so each coached worker becomes a partial coach in turn. What looked like an unaffordable luxury, a veteran at every elbow, becomes an affordable network, a culture in which interpreting signals out loud is simply how the floor works. Grace never hired a single full-time coach. She changed what it meant to be an experienced nurse on her floors, and the coaching came free with the change. The deepest version of the lever is not a program at all. It is a floor where the older hands ask the younger ones what they see before telling them what to do, where uncertainty is spoken rather than swallowed, and where the system's signal is treated as the beginning of a conversation rather than the end of one. A program can be cut in a downturn. A culture is much harder to lose, which is why the goal is always to convert the program into the culture as fast as you can.

The principle to carry out of the second lever is this. The algorithm can deliver the signal in an instant. Only a human can turn the signal into judgment, and judgment, not speed, is the thing you are actually paying your people for. A system that delivers signals faster while removing the people who turn signals into judgment has optimized the cheap half of the job and discarded the expensive half, which is the half that was the point.

Hybrid coaching reinstalls the human interpreter at the moment of feedback. But interpretation is only ever as good as the underlying skill of the person doing the interpreting, and the third tension, efficiency against learning, was precisely about that underlying skill quietly disappearing. So the next lever turns from the moment of feedback to the long maintenance of capability itself, and it belongs again to Priya, who had to work out how to keep her engineers able to do the very things she had so efficiently taught the machine to do for them.

> **THE EVIDENCE**
>
> Hybrid coaching is the second lever and the primary answer to cadence against safety. It reinstalls the interpretive step the system removes, the step between signal and reaction where a human weighs the signal in context, converting a reflex of blind obedience or blind dismissal back into judgment.
>
> It works on three of the catalogued harms at once: it restores psychological safety by separating the coach from the evaluator, rebuilds the metacognition the erosion removed by making the worker do the interpreting, and serves as an error-correction layer that catches the misreads the dashboard cannot see. Critically, the coach asks rather than tells, aiming to become unnecessary.

**MONDAY MORNING**

Pick your most experienced person and your most algorithm-anxious one, and pair them for thirty minutes a week with a single rule: the senior may only ask questions, never give answers, about the signals the system is throwing. Do not let the senior's manager hat into the room, and make sure it cannot touch the junior's score. In a month, ask the junior to walk you through how they read a flag. You will hear judgment where there used to be a reflex.

CHAPTER TEN

# Rotation and Explanation

*A skill you never practice is a skill you only think you have. A system you cannot question is one you only pretend to trust.*

Eighteen months after the outage at two in the morning, the payment service went down again, and this time Kofi fixed it.

He traced it himself, in about forty minutes, in a part of the system the tool would never have thought to examine, and he did not page anyone. What had changed was not Kofi's intelligence, which had never been in doubt, but his practice. For the past year he had spent one week in every six on deliberately hard problems with the tools turned down, and over that year the model in his head, the understanding of how the pieces of the system held each other up, had slowly been built. The building had not gotten any shorter. Priya was still shipping at the same pace. The foundation had simply gotten thicker, on purpose, one rotation at a time, and on the night it had to hold, it held.

Priya noticed something else that night, after the relief had worn off. Kofi had not just fixed the bug. He had been calm doing it, the specific calm of a person who knows they can find the thing, rather than the brittle confidence of one who only hopes the tool will. The rotation had not merely restored a skill. It had restored a relationship to the work, the sense that the system was his to understand rather than a mystery he serviced. That change in posture, from servicing the machine to commanding it, turned out to matter as much as the skill itself, and it was precisely the thing she had nearly optimized out of every engineer she had.

This chapter covers two levers rather than one, because they are two halves of a single idea. The third lever, rotational upskilling, keeps the human able to do what the system does. The fourth lever, explainable dashboards, keeps the human able to understand what the system is doing. One protects the skill. The other protects the comprehension. Apart they are useful. Together they let a person stand beside the machine as a peer rather than sit beneath it as a subject, and that standing-beside is the precondition for every other lever in this book actually working. A worker who cannot do the work and cannot understand the system has nothing to calibrate with, nothing to bring to coaching, and no real voice to exercise, because voice without competence is only noise.

## Lever Three: Rotational Upskilling

Rotational upskilling is the deliberate, scheduled practice of putting people back into the work the system has automated away, so that the underlying skill does not quietly vanish.

The pilot and the simulator are the cleanest model, so let us use them plainly. The autopilot flies the plane on every ordinary flight, and no one wants it otherwise, because it flies more smoothly and safely than a human hand. And yet the pilot still logs hours by hand in the simulator, regularly, throughout a career, not because the autopilot is unreliable but because the day it fails is the day you need a pilot who can actually fly, and that pilot does not exist unless someone kept them in practice during the long years when the autopilot made practice feel unnecessary. Rotational upskilling is the simulator, applied to any work an algorithm has taken over. It is the deliberate, scheduled, slightly inefficient maintenance of a capability you are not using today and cannot afford to be without on the day you need it.

It pays three returns. The first is that it keeps the model in the head alive, the deep structural understanding that Rosa had and Daniel lacked, that Lena had and Kofi was missing. The skill that erodes from disuse is maintained by deliberate use, and there is no other way to maintain it, because a skill is not a fact you can store on a shelf. It is a capacity that exists only in the using. The second return is that it keeps the senior pipeline stocked, defusing the generational time bomb from the efficiency chapter, because juniors who rotate through the hard work become the seniors who can one day supervise the tools. You are not just maintaining today's skill. You are manufacturing tomorrow's experts, on a ladder you had otherwise removed.

The third return is the least obvious and the most important for an operator who cares only about output. Rotation makes people better users of the tools. You cannot truly supervise a tool you could not do without, because you have no independent standard against which to judge its work. The engineer who has debugged by hand reviews the tool's code with real judgment, catching the confident error, because he knows what good looks like from the inside. The engineer who has never debugged by hand can only rubber-stamp, because he has no basis for doing anything else. Practice at the underlying skill is exactly what converts a tool-user into a tool-supervisor, and the difference between those two is the difference between an organization that commands its tools and one that is quietly commanded by them.

Building it well takes a few commitments. Make it scheduled and protected, a fixed fraction of time, the way Priya settled on one week in six, rather than something that happens when the backlog clears, because the backlog never clears and the rotation that waits for slack will wait forever. Frame it as training and not punishment, the simulator and not the penalty box, because the pilot framing removes the implication that needing practice means you are deficient. Choose the load-bearing skills, the ones that matter on the bad day, and rotate people specifically through the work the tool does for them, rather than through busywork that builds nothing. And make it count as real, visible, valued work, or it will be the very first thing cut the moment a deadline tightens, which is to say exactly when the future you are insuring against is being quietly mortgaged.

There are two guardrails. Do not over-rotate. You are buying insurance, not refusing the tool, and an operation that spends half its time with the tools turned off has simply surrendered the gains that justified them and will lose to a competitor who did not. And do not let it become theater, hours logged in a simulator that no one takes seriously, problems that are hard in appearance only. The rotation has to involve real, consequential, genuinely difficult work, the kind that can fail, because the struggle is the entire mechanism, and a struggle with nothing at stake teaches nothing at all.

## What Rotation Is Not

Because rotation is easy to caricature, it is worth saying clearly what it is not.

It is not nostalgia, a sentimental wish to do things the slow old way because the slow old way built character. The slow old way built skill, which is a different and entirely practical thing, and the point is the skill, not the slowness. It is not anti-tool. The tool ships the work on every ordinary day, and an operation that refuses the tool on principle will lose to one that does not, exactly as Priya's first attempt to slow everything down nearly did. And it is not an effort to make everyone a generalist who can do everything by hand, which would be both impossible and pointless. Rotation is targeted. It maintains the specific, load-bearing skills that matter on the bad day, the ones whose absence is catastrophic rather than merely inconvenient, and it deliberately ignores the rest.

Getting the target right is most of the craft. You are not trying to keep every skill alive against the tide of automation, which is a losing fight and not worth fighting. You are identifying the handful of capabilities your operation cannot survive without on its worst day, the debugging that saves the payment system, the clinical judgment that catches the patient the screen missed, the route knowledge that gets the truck across the river, and you are spending your rotation budget there and nowhere else. A rotation program that tries to preserve everything preserves nothing, because it spreads too thin to build real depth in anything. The discipline is in choosing the few skills genuinely worth the expense, and being ruthless about the rest.

Grace built her own version on the wards, and it looked nothing like an engineer's rotation, which is exactly the point. She had her nurses periodically work a stretch with the early-warning dashboard deliberately dimmed, relying on their own assessment first and checking the screen only afterward to compare. It was not that she distrusted the system. It was that she refused to let the assessment skill, the very thing that had made Ada able to feel bed nine going wrong, fade from disuse in a generation of nurses who had never once practiced without the screen. Same lever, different costume. The logistics version, the clinical version, and the engineering version all share a single spine: scheduled, protected practice in the load-bearing skill the system has taken over, framed as the maintenance of a reserve rather than a vote of no confidence in the tool.

## Lever Four: Explainable Dashboards

Rotation keeps the human able to do the work. The fourth lever keeps the human able to understand the system while they do it, and it is the direct answer to the trap Priya fell into when she opened the box and watched trust go down.

Recall the lesson of the transparency chapter. Disclosure is not understanding, and understanding is not trust. Dumping the model's weights on people is transparency of the useless kind, the kind that produces a glass box rather than a door. An explainable dashboard is not one that can be explained if someone digs hard enough. It is one that explains itself, in the moment, in a way a working person can actually use to weigh the signal against their own judgment. There is a structure for doing this well, and it is worth committing to memory, because most dashboards fail it on nearly every point. The structure spells the word CLEAR, which is what a good dashboard is and most are not.

C is for Context. Before anything else, the dashboard should answer why am I seeing this, and why now. A flag with no context is an alarm, and an alarm produces the reflex of obedience or dismissal we are trying to escape. A flag with context is information. This patient's risk score rose because three indicators moved together in the last twenty minutes tells the nurse what situation she is standing in, which is the first thing any judgment requires and the first thing most dashboards omit.

L is for Logic. The system should show how it reached its conclusion, in reasoning a human can follow, not raw weights and coefficients but the actual chain of inference. Rising heart rate, plus falling oxygen, plus a temperature trend, weighted toward the most recent change. The logic lets the worker check the system's reasoning against the reality in front of them, which is the thing a black box, however accurate, never permits, and the thing that turns a verdict into something a person can engage rather than merely receive.

E is for Evidence. The specific data behind the call, laid out so the worker can verify it against the world. This is what lets the human catch the misread. The nurse looks at the actual numbers and notices, as Bisi did, that the patient was just moved to a chair, which fully explains the heart rate the system found alarming. Evidence is the difference between a system that announces and a system that can be audited in the moment by the person closest to the truth, who is almost always better informed about this particular case than the model that scored it.

A is for Action, and this is the letter almost every dashboard omits, and the one that actually builds trust. The dashboard should tell the worker what they can do about the signal, including, above all, how to disagree with it. This is the recourse from the transparency chapter, the writable pane, the voice, built directly into the feedback itself rather than bolted on somewhere else. A flag that says here is what I see, and here is how you can tell me I am wrong is a flag a person can trust and improve. A flag that issues only a verdict is one they can only obey or resent, and we have seen at length where resentment leads.

R is for Recap. A short, plain summary the worker can act on and carry away, which closes the loop. Not the whole analysis, but the actionable distillation. Watch this patient, recheck in twenty minutes, escalate if it climbs. The recap converts understanding into a next step, which is the entire purpose of feedback in the first place. A dashboard that explains beautifully and leaves the worker unsure what to actually do has failed at the last yard, and a surprising number of sophisticated systems fail exactly there, mistaking the explanation for the point when the point was always the action.

Walk the five letters and you see what CLEAR really does. It converts a verdict into a conversation. Context and Logic and Evidence together give the worker enough to weigh the signal rather than blindly obey or ignore it, which is the very same interpretive step that hybrid coaching adds, now built into the dashboard itself so that it scales past the reach of any human coach. Action supplies the voice that the transparency chapter showed is the thing that actually builds trust, rather than the explanation everyone reaches for instead. And Recap closes the loop into a usable next step. A CLEAR dashboard is one a worker can think with. An ordinary dashboard is one they can only submit to, and submission, as every chapter of this book has shown, is where the competence quietly dies.

You do not need to rebuild your systems from scratch to do this. Start with the highest-stakes flags, the ones where a misread hurts someone, and make those CLEAR first. Then apply the only test that matters, which is to ask a worker to explain back what the dashboard told them and what they intend to do about it. If they cannot, the dashboard is not yet CLEAR, no matter how much logic you have exposed or how proud the vendor is of the explainability features. And wire the Action step to the writable panes and to the calibration room, so that a worker's disagreement with a flag becomes a signal that improves the system rather than a complaint that dies in a queue. That single wire connects the fourth lever back to the first and forward to the fifth, which is how you know you are building a loop and not a pile of separate tools.

## A Flag, Before and After

It helps to see a single flag rewritten through CLEAR, because the difference is larger than it sounds on paper.

Take a real alert from Marcus's world. Before CLEAR, a driver's tablet showed a red banner: route deviation detected, return to assigned route. That is a verdict and an order, and it produces exactly the two failures we have catalogued. The compliant driver obeys it even when he can see the assigned route is flooded, and the experienced driver learns to swipe it away as noise, because half the time the deviation was his own correct judgment. Either way the flag has taught the driver something corrosive about his own judgment, and it has done so several times a day, every day, for years.

After CLEAR, the same alert reads differently. Context: you have left the assigned route near the river crossing. Logic: the system prefers the assigned route because it was the fastest at dispatch. Evidence: here is the assigned route and your current position. Action: continue if you have information the system lacks, and tell us why with one tap, or return to route. Recap: deviation noted, your reason logged, drive safely. The same situation, the same data, and a completely different relationship between the driver and the system. The first version says obey me. The second says here is what I think and why, and you are the one actually on the road. One trains followers. The other trains judgment, and it does it in the four seconds a driver glances at a tablet, at a scale no human coach could ever reach.

## Why Most Dashboards Fail

Almost every dashboard in existence does the middle of CLEAR and skips the ends, and the pattern of what they skip is revealing.

They tend to do Logic, because logic is what engineers find interesting and what explainability features are sold on, and they sometimes do Evidence, because the data is right there. They reliably skip Context, because context requires understanding the worker's situation rather than the system's, and the system does not have that information unless someone deliberately gave it. They almost universally skip Action, because Action means building a real channel for disagreement, which costs the organization something and cedes a little control, and we have seen how hard leaders find that. And they skip Recap, because the designers, having explained everything, assume the worker can distill it themselves, forgetting that the worker is busy and the moment is brief.

The pattern is not random. Dashboards fail at exactly the two ends that serve the worker rather than the system. Context and Recap are about making the signal usable by a human under pressure, and Action is about giving that human a voice, and all three require the designer to take the worker's side rather than the system's. Logic and Evidence can be produced by pointing a camera at the model. Context, Action, and Recap can only be produced by caring what it is like to be the person receiving the flag, which is a harder and rarer thing to build into software, and exactly the thing that separates a dashboard people trust from one they merely endure.

There is a bonus in getting CLEAR right that operators tend to miss. A dashboard built with a real Action step does not only serve the worker. It makes the whole organization smarter, because every disagreement a worker registers is data about where the system and the world diverge, and that data, routed to the calibration room, is how the system learns. A verdict-only dashboard throws that information away. A CLEAR one harvests it. The explanation you build for the worker turns out to be the sensing apparatus the organization needs to keep its own systems honest, which means the most worker-friendly design is also, not by coincidence, the most institutionally intelligent one.

## Why the Two Belong Together

The two levers in this chapter are a matched pair, and seeing why is the key to the whole of Part Three.

Rotational upskilling keeps the human capable of doing what the machine does. Explainable dashboards keep the human capable of understanding what the machine is doing. One without the other fails in a specific and predictable way. A person who understands the dashboard perfectly but has lost the underlying skill can read the signal and still not act well on it, like a passenger who can follow the map and cannot drive the car. A person who has the skill but cannot understand the system fights it or ignores it, like a veteran driver who refuses to touch the app at all and loses its real benefits out of distrust. You need both at once, the capability to act and the comprehension to judge, and they reinforce each other, because the skill makes the explanation usable and the explanation makes the skill relevant. Together they produce the one thing this entire book is trying to protect: a human being who can stand beside the machine as a peer, neither its operator nor its subject.

Notice too that the two levers protect each other over time. Rotation without explanation produces people who can do the work but resent a system they cannot see into, and resentment slowly erodes the will to keep up the practice. Explanation without rotation produces people who understand the system beautifully and have no independent skill to weigh its output against, so the understanding has nothing to stand on. Run together, each gives the other something to hold. The skill makes the explanation meaningful, and the explanation makes the skill feel respected rather than nostalgic. This mutual reinforcement is the first clear hint of why the levers were designed as a loop rather than a list, a hint the final lever will make explicit.

The principle to carry out of the third and fourth levers is a single sentence with two clauses, because the levers are two clauses of one idea. Keep your people able to do what the machine does, and able to understand what the machine is doing, and you will always have someone who can take the wheel. Let either one atrophy and you have passengers, who are perfectly fine right up until the road does something the machine did not expect, which is the only moment any of this was ever about.

There is one more reason to take the peer-versus-subject distinction seriously, and it is about the worker rather than the work. People who stand beside the machine as peers tend to stay. People who sit beneath it as subjects leave, or worse, they stay and stop caring, which is the Carmen outcome and the most expensive one of all. The two levers in this chapter are, among everything else they do, a retention strategy and a dignity strategy, because a person who can do the work and understand the system experiences their job as a craft they are mastering, and a person who can do neither experiences it as a screen they obey. The first kind of worker compounds in value over a career. The second quietly leaves to become someone else's problem, taking with them whatever judgment they had left. Marcus had learned this the hard way with Carmen, and he had no intention of relearning it with a whole generation of Daniels. The levers were not charity. They were the cheapest insurance he had ever bought, and unlike most insurance, they paid a dividend every ordinary day, in the form of people who were getting better instead of getting hollow.

Four levers now stand built. Goal calibration gave people a voice over the targets. Hybrid coaching gave them an interpreter at the moment of feedback. Rotational upskilling kept their underlying skill alive. Explainable dashboards let them understand and contest the system in real time. Each has been a separate practice in a separate chapter, and each can stand more or less on its own. But they were never meant to stand alone. They were designed to connect, into a loop that feeds itself and grows stronger the longer it turns, and the fifth lever is the one that closes the circle and sets the whole thing in motion. It belongs to Priya, and it is where the three threads of this book finally come together into one.

### THE EVIDENCE

Levers three and four are a matched pair. Rotational upskilling answers efficiency against learning by scheduling deliberate practice in the work the system automates, which maintains the deep skill, restocks the senior pipeline, and turns tool-users into tool-supervisors who can catch the confident error.

Explainable dashboards answer transparency against trust through the CLEAR structure: Context, Logic, Evidence, Action, Recap. Context, Logic, and Evidence let a worker weigh the signal rather than obey or ignore it; Action supplies the recourse that actually builds trust; Recap closes the loop into a usable next step. The test of CLEAR is whether a worker can explain it back.

### MONDAY MORNING

Two moves this week. Schedule one block in which your team does, by hand, a thing the tool normally does for them, and protect that block like a customer meeting. Then take your single highest-stakes dashboard alert and run it through CLEAR: does it give the worker the context, the logic, the evidence, a clear action including how to disagree, and a one-line recap? Fix whichever letter is missing. It is almost always Action.

CHAPTER ELEVEN

# Voice Forums and the Loop

*A system that cannot hear its people will, in time, be run by people who have stopped speaking to it.*

The change that saved Priya's architecture did not come from an engineer, or a manager, or the tool. It came from a forum.

By this point Priya had built four of the five levers, and each was doing real work. But the thing that turned them from four separate practices into a single working system was the fifth, and it had started as the least promising of all of them. Once a month, a rotating group of engineers, from the newest juniors to the most senior principals, met with the people who owned the AI tooling and the people who owned the metrics, and they did exactly one thing. They decided, together, how the system that managed them would change. It was a small standing body with a strange amount of power, and Priya had very nearly not bothered to build it, because on paper it looked like one more meeting.

The win that changed her mind came from a junior engineer in the forum, perhaps a year into her career. She pointed out that the system rewarded merged pull requests, and so everyone optimized for merging fast, which meant that large, risky changes were being broken into many small merges that each looked perfectly safe and collectively were not. The metric was manufacturing the exact architectural fragility Priya had been fighting since the first outage at two in the morning, and nobody senior had seen it, because every individual merge passed every check. The junior saw it because she lived inside the incentive, and the incentive was shaping her hands every day. The forum changed how the system valued merges within the month, and a whole category of fragility began, slowly, to disappear.

This is the fifth lever, the moderated voice forum, and it is the one that closes the circle. The other four build the capacities of individuals: a voice over the target, an interpreter at the moment of feedback, a maintained skill, an understood system. The fifth takes all of those individual capacities and points them, collectively, at the system itself, so that the people who live inside the machine become the people who continuously redesign it. It is the lever that turns a set of managed individuals into a self-improving organization, and it is, not coincidentally, the hardest of the five to build and the easiest to do badly.

What moved Priya most was where the insight had come from. Not from her, the most senior engineer in the building. Not from the team that owned the metric, who thought about it constantly. Not from the tool, which had no idea it was doing harm. It had come from the person with the least tenure and the most direct exposure to the incentive, and it had surfaced only because there was a room where she could say it and be taken seriously. The intelligence had been in the organization the whole time, distributed across the people who lived inside the system. The forum had not created it. The forum had simply built a channel through which it could finally reach the place where the system was changed.

## What a Voice Forum Is

A voice forum is a moderated, standing channel through which the people who work under a system shape its ongoing design.

Every word of that is load-bearing, and the fastest way to understand it is by what it is not. It is not a suggestion box, which collects ideas and answers none of them, and so teaches people that speaking is pointless. It is not a town hall, which lets people vent and changes nothing, and so converts frustration into theater. It is not an engagement survey, which measures sentiment and acts on it quarterly if at all. It is a recurring body with real scope to change the system, moderated so that the changes are coherent rather than chaotic, with the leader still holding final authority. The forum proposes, shapes, and contests. The leader still decides. But what the forum decides about is not a narrow slice. Where calibration is about the targets specifically, the voice forum is about the whole system, the tool, the metrics, the dashboard, the rules, anything at all that shapes how the work is managed.

## The Forum and the Calibration Room

It is worth being precise about how the voice forum differs from the calibration room, because they look similar and do different jobs, and confusing them weakens both.

Calibration is narrow, frequent, and operational. It is about the targets, the specific metrics a particular team is measured against, and it runs often, close to the work, with the people that metric measures. The voice forum is broad, periodic, and structural. It is about the whole system, the tooling and the dashboards and the rules and the very logic of how the work is managed, and it runs less often, with a wider and more rotating set of people, including some who do not live under any single metric. Calibration tunes the instrument. The forum redesigns the orchestra. A team can calibrate its own targets every week and still need a forum to raise the question of whether the entire scoring approach is wrong, which is a question no single team's calibration can reach.

The two connect, of course, which is the point of a loop. A pattern that keeps surfacing in many teams' calibration rooms is exactly the kind of system-level issue the forum exists to address, and a change the forum makes to the system resets the targets that the calibration rooms then tune. The merged-pull-request problem was a forum problem precisely because it lived above any one team, in the system's basic definition of good work. No amount of local calibration would have caught it, because every team was correctly hitting a target that was collectively wrong. That is the forum's domain: the wrongness that is invisible from inside any single team and obvious only from above.

## Why It Comes Last

The voice forum is the fifth lever and not the first for a reason that took the other four to make clear. It requires all of them to already be working.

Think about what a genuinely useful voice forum demands of the people sitting in it. They need to believe their voice actually matters, which only calibration's track record of changed targets can establish, because people who have never seen their input change anything will not bring their real thoughts to a forum. They need to feel safe enough to contest the system out loud, in front of authority, which only the psychological safety that hybrid coaching rebuilds makes possible. They need the skill to have something worth saying about the work, which rotation preserves, because a workforce that no longer understands the work has no expert judgment to contribute. And they need to understand the system well enough to contest it intelligently rather than vaguely, which only explainable dashboards provide.

A voice forum built before those four capacities are in place is the empty writable pane in the high-power-distance office all over again, a door that no one walks through, a meeting that fills with silence or with the loudest voices and nothing in between. This is why so many well-meaning participation initiatives fail. They are the fifth lever attempted first, the capstone laid where the foundation should be. The forum is where the capacities the other four levers build finally get exercised together, on the system as a whole, and it cannot work until those capacities exist. Build it last, on top of the other four, and it works. Build it first, on top of nothing, and it becomes one more thing your people have learned to ignore.

## Who Sits in the Forum

The membership of a voice forum is a design decision, not an afterthought, and the wrong membership quietly determines the wrong outcome before anyone has said a word.

It should rotate, so that a fixed clique does not capture it and so that the experience of shaping the system spreads through the workforce rather than concentrating in a permanent committee. It should cross levels deliberately, putting the newest people in the same room as the most senior, because the junior who lives inside the incentive sees things the principal has long since stopped noticing, exactly as the merged-pull-request insight demonstrated. It should include the people who own the system, the metric owners and the tool owners, not as defendants but as participants, because a forum that excludes the builders produces proposals the builders will quietly ignore. And it must include a moderator whose loyalty is to the process rather than to any faction, which is the role that makes all the rest of it work.

Marcus learned, running his version, that the temptation is always to staff the forum with the articulate and the willing, and that this is a mistake, because the articulate and the willing are not where the hardest truths live. The hardest truths live with the people who have stopped speaking, the ones the spotlight taught to go quiet, and getting them into the room and getting them to talk is most of the moderator's job. A forum of the eager is a pleasant meeting that changes little. A forum that has done the work to include the reluctant is uncomfortable, and it is where the real redesign happens, because the reluctant are reluctant for reasons, and the reasons are usually the most important things anyone could tell you.

## Why Moderated Matters

The word moderated is carrying real weight in voice forum, and skipping it is the single most common way the lever fails in the hands of leaders who genuinely meant well.

Unmoderated voice does not become democracy. It becomes noise, or capture, or grievance. Noise, when too many unstructured opinions arrive at once and nothing coherent can be built from them. Capture, when the loudest or most senior voices dominate and the forum quietly becomes their instrument, the same failure that threatens the calibration room. Grievance, when the forum has no real scope to change anything and so curdles into a place where people come to complain and leave more bitter than they arrived. The moderator exists to prevent all three.

The moderator's job is not to control the conclusions but to structure the conversation. To keep the scope bounded and explicit, so the forum knows what it can change and what it cannot, and does not waste itself on the unchangeable. To synthesize many voices into coherent proposals a leader can actually act on. To make sure the quiet are heard over the loud, which is the same discipline that made Marcus's calibration room work. And above all to close the loop, reporting back at each session what changed and why, on everything that was raised at the last. An unmoderated forum produces a list of complaints. A moderated one produces a decision, and a decision that visibly changes the system is the thing that brings people back next month with their real thoughts instead of their safe ones.

## The Five-Lever Loop

Now the payoff that the whole of Part Three has been building toward, and the single most important idea in this half of the book. The five levers are not a list. They are a loop.

Walk it once, starting anywhere, because a loop has no true beginning. Calibration sets the right targets, the ones that measure the real goal rather than a convenient proxy for it. Explainable dashboards deliver the signals from those targets in a form people can actually understand and weigh. Hybrid coaching helps people interpret those signals in context and decide well. Rotational upskilling keeps people skilled enough to act on their decisions and to judge the system's output rather than merely accept it. And voice forums take everything the first four surface, the misread metrics, the confusing signals, the patterns the coaches notice, the gaps the skilled workers see, and feed all of it back into the design of the system, which resets the targets, which is calibration, which is where we came in. The end of the loop is the beginning of the loop. That is what makes it a loop and not a list.

And the loop has a property that no single lever has. It compounds. Each time around, the targets get truer, the signals get clearer, the interpretation gets sharper, the skills get deeper, and the system gets smarter, because the people inside it are continuously teaching it what it cannot see on its own. An organization running the loop does not merely hold steady against the competence paradox. It gets better at managing the paradox over time, because the loop is a learning system, and the specific thing it learns is how to keep its people growing inside a machine that was built to optimize them. The competence paradox is not defeated. It is harnessed, turned from a force that quietly degrades an organization into a flywheel that slowly improves it.

This is also why no single lever is ever enough, which is the mistake nearly every organization makes. They install one of the five, explainable dashboards perhaps, or a coaching program, and they are puzzled when it underperforms its promise. A lever pulled alone moves a little and then stalls, because the capacity it builds has nowhere to go. The five wired into a loop move the entire load, because each one covers the weaknesses of the others and feeds their strengths. Coaching without skill to coach toward fails; skill without voice to apply it curdles; voice without understanding is noise; understanding without trust is gamed; and trust, in the end, is built by calibration closing its loop. The loop is the unit of intervention. The lever is only a part, and a part installed alone behaves like a wheel without an axle, technically present and going nowhere.

## One Turn Around the Loop

It helps to watch a single complete turn of the loop in one organization, because described abstractly it can sound like a diagram rather than a thing that happens.

Start with the merged-pull-request problem in Priya's world. The voice forum surfaces it: a junior, made safe enough to speak by the coaching culture and skilled enough to see it by her rotations, raises that the merge metric is manufacturing fragility. The forum, moderated into a coherent proposal, sends it to calibration, where the target is recalibrated to value safe, coherent change over raw merge speed. The recalibrated target flows into the dashboard, which is rewritten through CLEAR so that engineers can see why a large change is now flagged for review and what they can do about it, including contest it. Hybrid coaching helps the engineers interpret the new signal rather than game it, and rotation keeps them skilled enough to make the architectural judgments the new metric now rewards.

And then the loop turns again. The new dashboard, with its Action step, starts collecting engineers' disagreements with the new flags, some of which reveal that the recalibrated metric is itself slightly wrong in a way no one anticipated, which surfaces in the next forum, which sends it back to calibration, and so on, around and around. Each turn the system gets a little truer and the people get a little sharper. No single turn solves anything permanently. The merge problem will mutate, a new incentive will bend in some new direction, and the loop will catch that too, because the loop is not aimed at any particular problem. It is aimed at the permanent condition of being managed by a system that cannot see everything, and that condition does not go away, so neither does the loop.

## When the Loop Breaks

Because the levers form a loop, a weakness in any one of them does not stay contained. It breaks the whole circuit, and learning to find the broken link is most of the work of running the system once it is built.

If calibration is weak, the targets stay wrong, and every downstream lever is faithfully serving a bad goal. If the dashboards are not CLEAR, people cannot understand the signals well enough to contribute useful disagreement, and the forum starves for want of good input. If coaching is missing, people are not safe enough to raise what they see, and the loop loses its richest source of signal. If rotation has lapsed, the workforce no longer has the skill to judge the system, and their contributions degrade into uninformed complaint. And if the forum is theater, everything the other four surface piles up with nowhere to go, and people stop feeding the loop because feeding it changed nothing. A broken loop usually presents as a single visible symptom, a forum that has gone quiet, a coaching program no one values, but the cause is often a different lever entirely, upstream, and the diagnostic skill is to trace the silence back to the link that actually failed.

This interdependence sounds like fragility, and in the building phase it is. But once the loop is turning, the same interdependence becomes resilience, because each lever reinforces the others and a brief weakness in one is buffered by the strength of the rest while you repair it. The loop is hard to start and hard to stop, which is exactly the property you want in a practice meant to run for the life of the organization. The danger is never that a turning loop suddenly collapses. The danger is that it slows, quarter by quarter, under the steady pressure to coast, until one day it has stopped and no one noticed, because a stopped loop and a slow one look identical for a surprisingly long time.

## Dynamic Equilibrium, Made Real

Cast your mind back to the second chapter, and the strange prescription that paradox theory offered. You do not solve a paradox. You manage it, toward something called a dynamic equilibrium. It sounded abstract then. The loop is what it looks like when it is real.

A dynamic equilibrium is not a resting point. It is a balance actively and continuously maintained, like a cyclist who stays upright only by steering constantly and minutely, correcting a thousand times a minute without ever thinking about it. The loop is that steering. The competence paradox never resolves, because the tension between the machine's gains and the human's growth is permanent, built into the very structure of algorithmic management, not a bug that a better version will one day fix. What the loop provides is not an end to the tension but a way to hold it productively, turn after turn, so that the gains keep arriving and the people keep growing, and neither pole is ever allowed to collapse the other. You never get to stop pedaling. But you do get to stay upright, and to go somewhere, which is a great deal more than the organizations that kept trying to fix the paradox ever managed, because they were looking for a kickstand on a machine that only works in motion.

## Where the Three Ended Up

By the time all five levers were turning, Marcus and Grace and Priya had each, separately, built a version of the same loop, in three industries that shared nothing but the shape of the problem.

And they had begun to borrow from one another, which was the thing I had quietly hoped for from the first morning I realized that their three calls were really one call. Marcus's calibration room became the model for Grace's metric reviews on the wards. Grace's protected coaching pause became the template for how Priya's seniors mentored juniors through the AI tools. Priya's writable panes became the design Marcus used to give his drivers a real voice inside the routing app. Three leaders, three industries, one loop, lent back and forth between them until none of them could quite remember whose idea each piece had originally been. That borrowing is the subject of the final chapter, because it points at something a good deal larger than any one of their three organizations.

What surprised me most about the three of them was how differently the same loop looked in each place. Marcus's forum met on a warehouse floor, in work boots, twenty minutes before a shift. Grace's met in a conference room between rounds, badges still on. Priya's met over video across three time zones. Different rhythms, different vocabularies, entirely different problems. And yet if you wrote down what each forum actually did, stripped of its industry, the three descriptions were identical. A standing body, made safe and skilled by the other four levers, redesigning the system that managed it, on a cadence, with the loop closing every time. The loop is industry-agnostic because the paradox is industry-agnostic. The costume changes. The body underneath is always the same.

There is a scaling property worth naming before we close. In a large organization the loop is not one loop but many, nested inside each other. Teams run small, fast loops on their own targets. Divisions run slower loops on their shared systems. The enterprise runs the slowest loop of all on the basic question of how it manages people at all. The loops feed one another the way the levers do, the small ones surfacing patterns the large ones act on. An organization that has built this well is not running an initiative. It is running a metabolism, a continuous process of sensing and adjusting that has become simply how the place works, as unremarkable and as essential as breathing, and as easy to take for granted right up until it stops. The leaders who build it eventually stop thinking of it as a thing they did and start thinking of it as a thing they are, which is the surest sign a loop has truly taken hold.

The principle to carry out of the fifth lever is the simplest of all of them, and the most easily ignored under pressure. Do not pull a lever. Build the loop. A fix ends, and the satisfaction of finishing it is exactly what makes it the wrong tool here. A loop turns, and only a turning thing can keep pace with a tension that never stops turning either. The organizations that will thrive in the age of the machine are not the ones that find the cleverest single intervention. They are the ones that build the humblest loop and keep it turning longest.

The five levers are the answer to the four tensions, and the loop is the answer to the fact that the tensions never end. But a loop is a machine, and a machine needs someone to run it, someone who understands why it matters and has the will to keep it turning when the quarterly pressure says stop and the visible numbers say it is safe to coast. That someone is the leader. Before we turn, in the final chapter, to what running this loop asks of the leader, there is the practical matter of how you actually put it to work: where to begin, how the work changes from one industry to another, how to measure the things the system cannot see, how the whole effort fails when it fails, and how to defend it to the people who govern by the first ledger alone. That practical work is what the next several chapters take up, before Marcus, Grace, and Priya sit down together in one room at last.

### THE EVIDENCE

Moderated voice forums are the fifth lever: a standing, moderated channel through which the people under a system shape its ongoing design, distinct from calibration in that its scope is the whole system rather than only the targets. It comes last because it requires the other four to function, the track record of changed targets, the safety to speak, the skill to contribute, and the understanding to contest.

The five levers form a loop, not a list. Calibration, explainable dashboards, hybrid coaching, rotational upskilling, and voice forums each feed the next and close back to the first. The loop compounds, getting better at managing the paradox over time, which is paradox theory's dynamic equilibrium made operational. The unit of intervention is the loop, not the single lever.

### MONDAY MORNING

Stand up one forum, once a month, with one rule that sets it apart from every meeting your people have already learned to ignore. It has the authority to change something real about the system that manages them, and the moderator must report back, at the next session, what actually changed because of the last. Start with a scope so small it is almost embarrassing. The size of the scope matters far less than the reliability of the loop closing, every time, without fail.

PART FOUR

# Putting It to Work

CHAPTER TWELVE

# The First Ninety Days

*People do not judge a change by what it promises. They judge it by what happens in the first month, and they decide faster than you think.*

I have watched more of these efforts die in the first ninety days than at any other point, and they nearly always die the same way.

The pattern is so common it deserves a name, because you will be tempted into every step of it. It begins with a leader who has understood the paradox, seen the levers, and felt the urgency, which is good, and then does the natural thing, which is to announce. There is an all-hands, a deck, a new platform, a memo with the word transformation somewhere in it. The leader describes all five levers, commits to all of them at once, and sets a date. The room nods. And then nothing closes.

By week three the calibration sessions have been scheduled and rescheduled. By week five the coaching pairs exist on a spreadsheet and have met exactly once. By week eight the forum has convened, generated a long list of complaints, and adjourned with no decision. By week ten the quarterly pressure arrives and the whole effort is quietly deprioritized, and by week twelve the people who briefly believed something might change have learned the more durable lesson, which is that it will not, and that the next leader who announces a transformation can be safely ignored.

That lesson is the most expensive thing a failed launch produces, and it outlasts the launch by years. A loop that is begun and abandoned does not return you to where you started. It leaves you worse off, because you have spent the one thing the whole system runs on, which is your people's belief that engaging honestly leads somewhere. This chapter is about how to spend the first ninety days so that you build that belief instead of burning it, because the difference is almost entirely in the first ninety days, and almost entirely under your control.

Sit with the cost of the failed launch a moment longer, because it is the part most leaders underweight when they are deciding how boldly to begin. An announcement that fails does not leave you where you started. It leaves you behind it, because your people now have evidence, fresh and specific, that engaging with your initiatives is a waste of their attention. The next leader who tries something real inherits that evidence and has to overcome it first. A failed launch is not a neutral event you can simply try again after; it is a withdrawal from an account that was already not full, and the account is the only thing the whole system runs on.

This is why the strategy of the first ninety days is built around a single closed loop rather than a broad rollout, and why the restraint it requires is strategic rather than timid. You are managing an account whose balance is your people's belief that honesty leads somewhere, and you cannot afford a withdrawal in the opening months when the balance is lowest. One small deposit that clears, one closed loop that everyone sees, is worth more at this stage than an ambitious program that risks a withdrawal, because early in the relationship the cost of a broken promise is at its highest and the benefit of a kept one compounds for the longest. Begin, therefore, as someone managing a scarce and precious balance, not as someone with credibility to spare, because you do not yet have any to spare.

## Why Ninety Days

The first ninety days matter out of all proportion to their length, because they are when your people decide whether you mean it, and they decide on evidence rather than on announcements.

An announcement costs you nothing, and so it signals nothing, because your people have seen announcements before and watched them evaporate. A closed loop, a thing that visibly changed because someone spoke, costs you something, and so it signals everything. The entire goal of the first ninety days is to produce one undeniable instance of a closed loop, witnessed by the people you most need to believe you. Everything else, the platform, the framework, the language, is secondary to that single piece of evidence.

This reframes the launch completely. You are not rolling out five levers in ninety days. You are trying to earn one piece of evidence, in front of the right audience, that voice in your organization now leads somewhere it did not lead before. Narrow and real beats broad and announced, every single time, because the paradox is fought with credibility, and credibility is built with evidence, not with scope. A leader who understands this will do far less in the first ninety days than their enthusiasm wants, and will succeed precisely because of that restraint.

## Before Day One

There is work to do before the ninety days even begin, and skipping it is how leaders find themselves committed in public to things they have not thought through.

Before you start, talk to people, quietly and without announcing anything. Find out what your people actually believe about whether speaking up changes anything, because that belief, more than any feature of your plan, determines what you are working with. If a prior effort failed here, you are starting from a deficit and need to know it. If the floor has simply never been asked, you are starting from neutral, which is easier. Either way, the temperature of that existing belief tells you how much credibility you have to spend and how carefully you have to spend it, and you cannot know it from your office.

And decide, before day one, what you will not promise. The instinct when you are excited about a change is to promise broadly, and broad promises are exactly what you cannot keep in ninety days, which means they are exactly what will sink you. Promise narrowly and specifically, or do not promise at all. It is far better to under-promise and let a closed loop overdeliver than to announce a transformation and let the gap between the announcement and the reality teach the lesson you are trying to avoid. The first ninety days are governed by a single rule that starts before they do: commit only to what you will visibly deliver, and deliver it.

## Days One to Thirty: Diagnose and Choose

Spend the first month doing two things well and resisting the strong urge to do more.

First, diagnose. Use the questions in the back of this book, or your own, to find which of the four tensions is doing the most damage to your organization right now. Talk to people, and not only the articulate and the willing. Read the second ledger you have probably been ignoring. Do not guess, because the wrong diagnosis sends you confidently to the wrong lever and wastes the entire window. The discipline of the first thirty days is to find the real tension rather than the one you already had a solution for.

Second, choose one lever and one place to start. One. The temptation to do all five at once is the single most reliable way to accomplish none of them, because five half-built levers close no loops, and the closing of a loop is the only thing that matters in this window. Pick the lever that answers your worst tension, pick the smallest real scope for it, one team, one metric, one shift, and commit to that and nothing else. You are not being timid. You are being strategic about where to spend a credibility you do not yet have and cannot afford to waste.

There is a third thing to settle quietly in the first month: who sponsors this, and whether they will hold. A loop dies fast without a sponsor senior enough to protect the time it costs against the quarterly pressure that is certainly coming for it. If you are that sponsor, decide now, honestly, whether you will hold the line when the quarter gets hard. If you are not senior enough to be it, secure one before you start, because a launch the sponsor abandons in week ten is worse than a launch that never happened, and the sponsor who will not hold is the most common hidden cause of the failures I have watched.

## Choosing the Right First Lever

The choice of which lever to build first is not arbitrary, and getting it right is most of what determines whether the first loop closes cleanly.

Match the lever to the symptom. If your best people are leaving and your scores are rising, the tension is specificity against autonomy, and you start with goal calibration, because it produces a visible, fast loop closing: a metric changes in a room because people spoke, and everyone sees it. If your floor has gone quiet, the tension is cadence against safety, and you start with hybrid coaching, though its loop closes more slowly and is harder to show, which matters for a first lever. If your bench is thinning, the tension is efficiency against learning, and you start with rotation, accepting that its payoff is the slowest of all to demonstrate.

Notice that the levers differ in how quickly they produce visible evidence, and in the first ninety days that property matters as much as the underlying fit. Calibration closes a loop you can show in a single session, which makes it an unusually good first lever even when it is not the perfect fit, because the whole game is producing visible evidence fast. Rotation, by contrast, is the right answer to a real and serious tension and a poor first lever, because its evidence arrives in quarters, not weeks, and the first ninety days cannot wait quarters. When the perfectly-fitted lever is slow to show results, it is often wiser to start with a faster-closing one to build credibility, and reach the slow one second, on the trust the first one earned.

Whatever you choose, choose one. The single discipline that matters more than which lever you pick is that you pick one and resist the rest until it has closed a loop. A leader who cannot bear to start narrow, who insists on launching three levers because all three tensions are real, will close none of them and learn nothing about which would have worked, because three half-built levers produce three half-closed loops, and a half-closed loop teaches your people exactly what a fully open one does.

## The Sponsor Who Holds

Every surviving loop I have seen had a sponsor senior enough to protect it and stubborn enough to keep protecting it when the quarter turned hard. Every dead one was missing that person at the moment it mattered.

The sponsor's job is narrow and decisive. It is to hold the line on the one or two commitments that the quarterly pressure will come for first, the protected coaching time, the rotation block, the inconvenient metric change the calibration room produced. These are the things that look optional under deadline pressure and are not, and someone with enough authority has to say so and mean it when the pressure arrives, which it will, on schedule, around week ten. A loop without that person survives until the first hard quarter and then quietly dies, and everyone learns the lesson again.

If you have the authority to be that sponsor, the question is not whether you can but whether you will, honestly, when holding the line costs you something visible in the quarter. Decide that now, in the cold, because you will not decide it well in the heat. And if you do not have the authority, do not start until you have recruited someone who does and who has made the same cold decision, because a launch that depends on a sponsor who folds is worse than no launch, and the most common way these efforts die is not bad design but a sponsor who meant well in the calm and could not hold in the storm.

## Days Thirty-One to Sixty: Close One Loop

The middle month is for running the lever once and closing its loop in front of everyone who needs to see it.

If you chose calibration, you run the session, you change a metric in the room, and you report the change. If you chose coaching, you pair people, you protect the time fiercely, and you let the first interpreted signals actually happen. The specific lever matters far less than the closing. The move that earns the ninety days is small and visible: something the system did changed because a person spoke, and everyone can see both halves at once, the speaking and the change that followed it. That is the whole of it. One visible cause, one visible effect, with a human in between.

Protect this with everything you have, because everything will conspire against it. The change will be inconvenient. Someone reasonable will want to defer it. The quarterly number will demand your attention at exactly the wrong moment. And if you let the first loop fail to close, on any excuse however good, you will have taught the precise lesson you were trying to avoid, that voice here leads nowhere after all. In the first sixty days, closing one loop is more important than the quarter, because it is the investment the quarter is quietly borrowing against, and a quarter that eats its own seed corn is not the win it appears to be.

## Days Sixty-One to Ninety: Prove and Widen

The last month is for making the closed loop undeniable, and for using it to recruit the people you will need next.

Show the change and its effect, in whatever terms your people actually care about, not in the terms your framework prefers. Find the skeptic who watched the loop close, the one who expected it to be theater, and ask them what they saw, because a convert who began as a skeptic is worth ten enthusiasts when it comes to persuading the rest. And only now, with one loop genuinely proven, begin the second lever, built on the trust the first one earned. The sequence matters: prove, then widen, never the reverse, because widening on the strength of an unproven promise is just a larger version of the announcement that fails.

Resist widening too fast even here, because the failure mode of a successful first loop is triumphant over-expansion, the leader who declares victory and scales to the whole organization before the practice is robust enough to survive the scaling. Grow at the speed of proven loops, not at the speed of your own enthusiasm. Each new loop you add should be as real as the first one was, because the moment any single loop becomes theater, the credibility you spent ninety days building begins quietly draining back out, and it drains faster than it filled.

## The Witnessed Loop

The phrase witnessed loop deserves unpacking, because the word witnessed is doing as much work as the word loop, and most leaders who fail at this get the loop right and the witnessing wrong.

It is not enough for a loop to close. It has to close where the people you most need to convince can see both halves of it, the speaking and the change that followed, and can connect the one to the other. A change made in a back room and announced later is not a witnessed loop. It is an announcement, and your people, having learned to discount announcements, will discount it too, no matter how real the change behind it was. The connection between the voice and the consequence has to be visible, immediate, and undeniable, or the loop closes in private and teaches no one anything.

This is the deeper reason the calibration room makes such a strong first lever. The closing happens in front of everyone, the metric changing in the same hour someone argued for it, the cause and effect compressed into a single visible moment that no one can miss or reinterpret. Compare a change made quietly by email three weeks after a session, which your people cannot reliably connect to anything they said, and which therefore signals nothing about whether their voice mattered. Same change, same loop, and one builds credibility while the other squanders it, entirely on the question of whether the closing was witnessed.

So choose not merely a loop that will close but one that will close visibly, in front of the specific people whose belief you most need, and then stage the witnessing on purpose. Name what was raised, name what changed because of it, and name the connection out loud, so that no one has to infer it. This feels almost theatrical to a leader who would rather just make the change and move on, and the instinct to be quietly efficient about it is exactly the instinct to resist, because half the value of the first closed loop is in the watching. A loop that closes and is not witnessed has done its operational job and wasted its credibility-building one, which in the first ninety days is the job that matters more.

There is a multiplier here worth naming. A single well-witnessed loop does not convince only the people in the room. It travels, because the people who saw it tell the people who did not, and a story about a change that actually happened because someone spoke is the most persuasive thing in your organization, far more persuasive than anything you could say from a stage. The witnessed loop is, in effect, a story you are manufacturing on purpose, the founding story of the belief you are trying to build, and a leader who understands that will spend real care making the first one clean, visible, and easy to retell, because it will be retold, and the retelling is how a single closed loop becomes a culture that expects loops to close.

## What to Measure in Ninety Days

Do not measure the things that take a year to move, and then despair when ninety days have not moved them.

Turnover, safety rates, codebase health, these are lagging indicators, real but slow, and judging a ninety-day launch by them is like judging a diet by the bathroom scale on the third morning. The numbers that matter in this window are the leading indicators, the ones that move in weeks. Are people voicing more concerns than they did. Are they filing more contests, flagging more scores, raising more in the forum. Is the quiet floor getting louder in the specific way that means people have started to believe it is safe to speak.

The single best ninety-day measure is participation that costs the participant something. A worker contesting a metric, a nurse flagging a score she will have to defend, an engineer annotating a system in writing with his name attached. Those behaviors are the early and honest signal that your people believe voice is real, and they move long before any lagging number does. If costly participation is rising, the loop is working, whatever the slow numbers say so far. If it is flat after sixty days, no lagging improvement is coming, and you should find out why now, while you still have most of the window left to fix it.

## Inheriting a Poisoned Well

Many leaders do not start from neutral. They start in the wreckage of a previous effort that promised much and delivered nothing, and that starting point changes the ninety days completely.

If your people have been here before, asked for their input and then ignored, they will meet your launch with a skepticism you have not earned and cannot argue away, because it was earned for you by whoever came before. The mistake is to treat their skepticism as an obstacle to overcome with enthusiasm, because enthusiasm is exactly what the last failed effort had, and your people learned to discount it. The only thing that works on a poisoned well is the slow, unglamorous accumulation of kept promises, starting smaller than you would otherwise dare and proving each one before you make the next.

So when you inherit a poisoned well, shrink the first ninety days even further. Promise almost nothing. Pick a loop so small that closing it is almost trivial, and close it visibly, and then do it again, and let the evidence rebuild what words cannot. Name the past failure honestly if it is known to everyone, because pretending it did not happen confirms their suspicion that you are the same as the last one. The recovery is slower than a clean start, sometimes by a wide margin, but it follows the same logic, only with a steeper hill at the front: credibility is built one kept promise at a time, and a well that was poisoned simply needs more of them before anyone will drink.

## The Three Ways It Dies

Launches die three ways, and all three are avoidable once you can see them coming.

They die of over-scope, when a leader builds five shallow levers instead of one real one and closes no loop at all. They die of no follow-through, when the first loop is allowed not to close on some reasonable-sounding excuse, teaching the fatal lesson in a single stroke. And they die of the wrong sponsor, when the person protecting the effort turns out not to be senior enough, or not committed enough, to hold the line against the quarter when the quarter finally pushes. Each of the three has killed more good intentions than bad ideas ever have.

Notice that none of the three is about the levers themselves. The mechanics are not what kills launches. Credibility is, and credibility is a function of scope kept small, loops kept closing, and a sponsor who holds. Get those three right and the worst that happens to you is slow progress, which is fine, because slow progress compounds. Get any one of them wrong and you will spend your people's belief and leave the place harder to lead than you found it, which is the one outcome worse than having done nothing at all.

None of this is complicated, which is part of why it is so often gotten wrong. The first ninety days do not reward cleverness, ambition, or a sophisticated rollout plan. They reward restraint, follow-through, and the patience to let one small kept promise do its quiet work, which are unglamorous virtues that a leader eager to make a mark finds genuinely hard to choose. The leaders who succeed in this window are rarely the boldest ones. They are the ones disciplined enough to do far less than they could, and to do that little thing completely, and to trust that the credibility it builds will carry the rest.

The principle to carry out of the first ninety days is to stop trying to be impressive and start trying to be believed. One closed loop, witnessed by the right people, is worth more than five levers announced to a full room, because the paradox is won on credibility, and credibility is built one kept promise at a time, beginning with the very first one you make.

The ninety-day arc gets you started wherever you stand. But the shape of the work changes depending on where exactly that is, because the paradox, identical underneath, wears a different disguise in a warehouse than in a hospital than in a software team, and the disguise is what fools most leaders into thinking their situation is unique. So the next chapter takes the framework into specific industries, and shows how to find your own dominant tension and your own load-bearing skill, wherever it is that you happen to work.

### THE EVIDENCE

The first ninety days are won by producing one closed loop, witnessed by the people you most need to believe you, not by announcing five levers. Credibility, not scope, is the binding constraint, and it is built with evidence that voice now leads somewhere it did not before.

Measure leading indicators that move in weeks, especially costly participation: workers contesting metrics, flagging scores, annotating systems with their names attached. These signal belief long before lagging numbers like turnover move. Launches die three ways: over-scope, no follow-through, and a sponsor who will not hold against the quarter.

### MONDAY MORNING

This week, do not launch anything. Instead, pick the one tension hurting you most, the one lever that answers it, and the smallest real place to try it, one team and one metric. Write down the single loop you will close in the next sixty days and the date you will report the change to the people it affects. Tell no one it is a transformation. Just close the loop, and let them notice on their own.

CHAPTER THIRTEEN

# Industry Playbooks

*The paradox is the same everywhere. The disguise it wears is not, and the disguise is what fools you into thinking your situation is unique.*

When Marcus, Grace, and Priya finally compared their loops in detail, the thing that struck all three was how little their solutions looked alike, and how completely they were the same underneath.

Marcus calibrated routing metrics with drivers in work boots before a shift. Grace ran protected coaching pauses with nurses between rounds. Priya held voice forums with engineers over video across three time zones. If you had watched any two of these without knowing the framework, you would have concluded they were solving unrelated problems in unrelated ways. And if you had written down what each was actually doing, stripped of its industry, you would have found a single structure repeated three times. That gap, between how different the surface looks and how identical the structure is, is the subject of this chapter, because it is the thing that most often convinces a leader that the framework cannot apply to them.

It always can. But applying it well requires knowing what travels unchanged from one industry to another and what has to be rebuilt locally from scratch, and getting that distinction wrong is how leaders either dismiss the framework as not fitting their world or import it so literally that it fails. So before the playbooks, the distinction itself.

## What Travels and What Stays Home

The framework travels completely. The specifics travel not at all, and pretending otherwise in either direction is the mistake.

What travels is the structure. The four tensions are present wherever a system optimizes a person, because they arise from the basic facts of human autonomy, safety, learning, and trust, which do not change between a warehouse and a hospital. The five levers answer those tensions wherever they appear, because they are built on those same human facts. And the loop is the loop everywhere, because a permanent tension everywhere requires a continuous practice. None of that needs translation. It is as true in a call center as in a cockpit.

What stays home is everything specific. Which tension dominates, the exact cadence at which feedback turns corrosive, which skills are load-bearing and must be protected by rotation, who can serve as a coach, how much power distance you are working against, and what the system is even optimizing. All of that is local, and it must be discovered in your own organization rather than imported from a book or a competitor. The error of the dismissive leader is to see that the specifics differ and conclude the framework does not apply. The error of the eager leader is to copy another industry's specifics literally. The craft is to carry the structure across unchanged and rebuild the specifics on arrival.

## Why the Disguise Fools Us

Before the playbooks, it is worth understanding why the disguise works so well, because the same psychology that fools other leaders is waiting to fool you about your own situation.

Every leader believes their situation is special, and in the ways they can see, they are right. The warehouse really is nothing like the ward. The cadence of a trading floor really is nothing like the cadence of a classroom. The specific pressures, vocabularies, regulations, and rhythms of any given workplace are genuinely particular, and a leader immersed in them all day sees the particularity far more vividly than the underlying commonality, because the particularity is what they wrestle with and the commonality is invisible by its nature. The disguise is not a trick someone plays. It is the natural result of standing close to one instance of a general pattern and seeing the instance in high resolution and the pattern not at all.

This is why the framework so often gets dismissed by exactly the people it would help most. A logistics executive reads about a hospital and concludes, reasonably, that nurses and drivers have nothing in common, missing that the thing they have in common is not their work but their relationship to an optimizing system, which is identical. The dismissal feels like hard-headed realism, a refusal to be taken in by a one-size-fits-all theory, and it is actually the opposite, a failure to see past the costume to the body underneath. The leaders who benefit are the ones who can hold both truths at once: my situation is genuinely particular in its specifics, and it is genuinely universal in its structure, and I must respect both or I will either dismiss the framework or misapply it.

## The Universal Spine

Underneath every playbook in this chapter runs the same spine, and naming it explicitly makes the per-industry versions easier to build and adapt.

The spine is this. Find the dominant tension from the shape of the work, because the shape of the work determines which of the four human needs the system most threatens. Lead with the lever that answers that tension, because a lever aimed at the wrong tension wastes the credibility you need. Protect the load-bearing skill through rotation, because every industry has one capability it cannot survive without on its worst day. Build voice with real recourse, because every system, in every field, is a black box to someone until you give them a door. And wire it all into a loop, because the tension never ends in any industry, so the practice cannot end either. That spine does not vary. Only the flesh on it does.

Once you can see the spine, the playbooks below stop being separate prescriptions and become five instances of one prescription, which is the only way to use them rightly. You are not looking for the playbook that matches your industry, because the perfect match may not be among them. You are looking at how each leader found their dominant tension, named their load-bearing skill, and sequenced their levers, so that you can run the same procedure in your own field, including a field no one has written a playbook for yet. The playbooks are worked examples of a method, and the method, not the examples, is the thing to carry away, because the method is what travels to wherever you actually work.

## The Logistics Playbook

In logistics and field operations, the dominant tensions are specificity against autonomy and cadence against safety, and they arrive together.

The work is physical, distributed, and increasingly directed turn by turn by routing and dispatch systems that issue precise, fast-refreshing instructions to people who can see things the system cannot, the flooded bridge, the unsafe dock, the fragile load. The autonomy erosion is acute because the guidance is so specific, and the safety risk is acute because the cadence is so high and the work has physical consequences. The load-bearing skill to protect through rotation is situational judgment, the route knowledge and the read of conditions that lets a driver override the system correctly on the rare day it is wrong.

The playbook follows. Calibrate the routing and scoring metrics with the drivers, in standing sessions, because they are the only ones who can see what the metric punishes by mistake, as Marcus learned with the consolidation step and the fragile shipments. Give them a writable pane inside the routing app so a deviation can carry a logged reason rather than a silent infraction. Protect situational judgment with periodic work in harder conditions rather than letting the system carry every easy day. And run a forum that can change the dispatch logic itself, because the people on the road are the richest sensor the system has and the one it most reliably ignores.

## The Healthcare Playbook

In clinical work, the dominant tensions are cadence against safety and transparency against trust, sharpened by stakes that do not forgive error.

Early-warning systems and clinical decision support issue rapid signals to people whose work can kill or save, which makes psychological safety a literal matter of patient survival, as Grace learned at bed nine. The spotlight effect is severe because the monitoring is constant and the consequences of a wrong move are catastrophic, so the quiet that high opaque cadence produces is more dangerous here than anywhere. The load-bearing skill is clinical assessment, the ability to read a patient directly, which fades quietly in a generation of clinicians who have only ever worked with the screen between them and the bedside.

The playbook. Lead with hybrid coaching, because the protected, off-the-record pause is what restores the safety the cadence destroys, and because clinical judgment is built by interpreting signals with a mentor rather than obeying them alone. Build explainable signals with real recourse, the CLEAR structure, so a nurse can contest a score and add the context the system missed, which both improves the system and rebuilds the trust that constant alerting erodes. Rotate clinicians through assessment with the dashboard dimmed, the way Grace did, so the reserve skill survives. And handle power distance deliberately, because medicine's steep hierarchies mean a writable pane sits unused until senior clinicians visibly invite and act on dissent from junior ones.

## The Software Playbook

In software and knowledge work, the dominant tension is efficiency against learning, the taller building on the thinner foundation.

AI coding tools and automated review let teams ship faster than ever while quietly removing the struggle through which deep engineering skill was always built, producing Priya's exact problem: juniors who are productive immediately and cannot debug the system at two in the morning, and a senior pipeline that is no longer being stocked because no one is doing the hard work that used to make seniors. The load-bearing skill is the deep structural understanding of the system, the model in the head that lets a person find the bug the tool wrote and cannot see.

The playbook. Lead with rotational upskilling, because the skill erosion is the primary harm here and the simulator logic is the direct answer, scheduled practice on hard problems with the tools turned down. Make the AI tooling explainable and contestable, since engineers who cannot see or argue with the system that scores them will game it expertly. Run forums with real authority over the tooling and the metrics, because engineers live inside incentives they can see clearly and will surface the merged-pull-request kind of fragility no manager can. And calibrate what good work even means, because in knowledge work the metric is especially prone to rewarding visible speed over invisible quality.

## Common Mistakes, by Industry

Each industry has a characteristic way of getting the framework wrong, usually by reaching for the lever that feels natural to its culture rather than the one its dominant tension requires.

Logistics operators tend to over-rely on calibration and skip the writable pane, because adjusting metrics feels like managing and giving drivers a live voice inside the app feels like losing control of the route. The result is a calibration room that improves the targets while the daily experience of being overridden by the system continues unchanged, and the autonomy erosion persists because the lever that addressed the moment-to-moment experience was the one that got cut. The fix is to remember that calibration sets the target and the writable pane handles the moment, and the moment is where the driver actually lives.

Healthcare organizations tend to over-invest in explanation and under-invest in voice, because clinical culture trusts documentation and protocol, so the instinct is to explain the system thoroughly and consider trust handled. This produces Lena's glass box in scrubs: clinicians who understand the early-warning system completely and trust it no more for the understanding, because explanation without the writable pane and the protected coaching pause answers the wrong question. The fix is to lead with coaching and recourse, not with documentation, however much documentation suits the culture.

Software organizations tend to over-build the explainable dashboard and neglect rotation, because engineers love a well-instrumented, transparent system and find scheduled practice on hard problems with the tools off to be an affront to efficiency. The result is a beautifully explainable tool supervised by engineers who are slowly losing the skill to supervise it, the taller building with the ever-thinner foundation, now well-documented. The fix is to accept the slow, unglamorous insurance of rotation as the primary lever, because the dominant tension in knowledge work is the skill erosion, and no amount of explainability addresses that.

## The Finance Playbook

In finance, the dominant tension is transparency against trust, because opaque models increasingly drive decisions that the people executing them, and the people affected by them, cannot interrogate.

Risk models, credit decisions, trading signals, and compliance flags arrive as outputs from systems whose logic is proprietary, complex, or both, to be acted on by professionals whose own judgment is considerable and increasingly sidelined. The trust problem is acute because the stakes are high and the recourse is usually nonexistent, and the gaming problem follows, because professionals who distrust a model they cannot contest will find ways to work around it that the model cannot see. The load-bearing skill is independent analytical judgment, the ability to evaluate whether a model's output makes sense, which atrophies when professionals are reduced to executing outputs they are not permitted to question.

The playbook leads with explainable dashboards and voice forums. Make the models' reasoning legible to the people who must act on them, with a real Action step so a professional can flag an output that contradicts their judgment and have the flag matter. Run forums with authority over how the models are used and weighted, because the professionals can see failure modes the model builders cannot. And protect analytical judgment through rotation that keeps people building and stress-testing their own analyses rather than only consuming the model's, because a finance function that can only execute a model it does not understand is one model error away from a catastrophe no one on the floor was equipped to catch.

## Education, Retail, and the Public Sector

Three more fields show how to locate the dominant tension quickly, because each one's tension is driven by the basic shape of its work.

In education, adaptive and analytics-driven systems optimize for measurable gains, so the dominant tension is efficiency against learning, the same one software faces, only with students rather than engineers as the people being optimized. The lever set leads with protecting the slower, harder, less measurable development that the systems crowd out, and with calibrating what the metrics reward, because a school that optimizes test-measurable gains will quietly defund everything it cannot measure. In retail and the gig economy, scheduling and task systems direct workers minutely and at high cadence, so specificity against autonomy and cadence against safety dominate, exactly as in logistics, and the playbook resembles the logistics one: calibration, the writable pane, and protected judgment.

In the public sector, transparency against trust dominates above all, because the people the systems judge are citizens with no exit and every reason to distrust a verdict they cannot contest, and the professionals administering the systems are caught between a public that distrusts the algorithm and an institution that defers to it. The playbook leads with explainability and recourse, both for the workers administering the systems and, harder and more important, for the citizens subject to them, because a public system that cannot be contested is the purest form of the glass box, and its costs land not on a balance sheet but on the legitimacy of the institution itself. In every one of these fields, the move is the same: find the dominant tension from the shape of the work, and lead with the lever that answers it.

## What the Three Leaders Borrowed

The most useful thing the three leaders did was steal from one another across the boundaries of their industries, and the borrowing always worked because they took structure rather than specifics.

Marcus took Grace's protected coaching pause, designed for nurses, and adapted it for drivers debriefing a hard route, keeping the structure, an off-the-record conversation with a non-evaluating peer, and rebuilding the specifics around logistics. Grace took Priya's writable panes, designed for engineers contesting code scores, and built them into her clinical dashboards so nurses could annotate a flag, again keeping the structure and rebuilding the specifics. Priya took Marcus's calibration room, designed for warehouse metrics, and ran it on her engineering targets. None of these transfers would have worked as literal copies, because a driver is not a nurse is not an engineer, but every one worked as a structural transfer, because the structure was built from what the three workers shared as human beings under optimizing systems.

This is the practical payoff of understanding what travels. Once you can separate the structure of a lever from its industry-specific clothing, the entire body of practice from every other field becomes available to you, to be adapted rather than copied. The leader who can do this is not limited to the playbook of their own industry; they can raid all of them, taking the structural innovation from healthcare or software or logistics and rebuilding its specifics for their own world. The four tensions make every industry's hard-won lessons relevant to every other, which is why a book that ranges across warehouses and wards and codebases is not a collection of unrelated cases but a single set of moves shown in many costumes. The leader who internalizes that stops asking whether a lesson from another industry applies to them, and starts asking only how to rebuild its specifics for their own ground, which is a far more useful question and the one this chapter exists to teach.

## Where the Thresholds Differ

One specific that varies sharply across industries, and that leaders most often import wrongly, is the threshold at which feedback turns from helpful to corrosive.

The propositions in this book name particular numbers, prompts exceeding thirty an hour, nudges of twenty or more a shift, a refresh every few minutes, and those numbers are real findings from the studies, but they are not universal constants you can paste onto your own operation. They are the thresholds observed in the settings that were studied, and your setting may turn corrosive sooner or tolerate more, depending on the nature of the work and the people. High-stakes, high-judgment work, a clinician or a pilot, turns corrosive at a lower cadence, because the spotlight effect is fiercer where the consequences are graver. Routine, low-stakes work may tolerate a faster cadence before the same harm appears.

So treat the published thresholds as starting hypotheses to test in your own context rather than settings to dial in. Watch for the symptoms the thresholds predict, the autonomy loss, the falling voice, the crowded-out exploration, and find the cadence at which they actually begin for your people doing your work, which is the only threshold that governs your decisions. A leader who imports the number thirty and applies it literally to a context where harm begins at fifteen will reassure themselves with a cadence that is already corroding their floor, and a leader who applies it to a context that tolerates sixty will throttle a system needlessly. The threshold travels as a concept, the point at which feedback turns corrosive, and stays home as a number, which has to be discovered where you are.

This is the per-industry specific in miniature, and it stands for all the others. The structure, that there is a corrosive threshold, is universal and you can trust it. The number, where the threshold sits, is local and you must find it. Confuse the two, in either direction, and you will either dismiss a real finding because your number differs from the book's, or apply the book's number where it does not belong and miss the harm it was meant to warn you of. The whole craft of the playbook is this discipline applied across every specific at once: carry the structure, discover the numbers, and never mistake one for the other.

## Finding Your Own Playbook

If your industry is not one of these three, you are in the majority, and the method for building your own playbook is the same one the three leaders used without naming it.

Start by finding your dominant tension, using the diagnostic in the appendix. In finance, it is often transparency against trust, as opaque models drive decisions people cannot interrogate. In retail and the gig economy, it is frequently specificity against autonomy and cadence against safety, as scheduling and task systems direct workers minutely. In education, it tends toward efficiency against learning, as adaptive systems optimize for measurable gains and risk crowding out the slower development they cannot see. In the public sector, transparency against trust dominates, because the people the systems judge are citizens with no exit and every reason to distrust a verdict they cannot contest.

Then find your load-bearing skill, the capability your operation cannot survive without on its worst day, and your answer to who can coach and how steep your hierarchy runs. With those four answers, the dominant tension, the load-bearing skill, the available coaches, and the power distance, you can assemble your own playbook from the five levers, leading with the one that answers your dominant tension and sequencing the rest behind it. The framework does not have to be adapted, because it was built from the human facts your industry shares with all the others. Only the specifics have to be discovered, and they can only be discovered where you are, which is the one place the book cannot reach and you can.

There is a quiet freedom in understanding that the paradox is universal, and it is worth naming for the leader who feels alone with their version of it. If the structure is shared across every industry, then you are never the first to face your problem, however particular its costume. Someone in healthcare has wrestled with a version of your autonomy erosion; someone in logistics has fought a version of your trust deficit; someone in software has watched a version of your skills fade. Their structural solutions are available to you, to be adapted rather than copied, which means the accumulated practice of every field becomes a resource for yours. The universality that makes the paradox inescapable is the same universality that means you never have to solve it from scratch.

The principle to carry out of the playbooks is to trust the structure and distrust the borrowed specifics. The paradox is universal, which is why the framework travels, and its disguise is local, which is why you must build the details yourself. A leader who copies another industry's playbook fails; a leader who copies its structure and rebuilds its specifics succeeds, and the difference between the two is the whole of the craft of applying this anywhere new.

Knowing which tension dominates and which skill is load-bearing tells you what to protect. But protecting something means, eventually, knowing whether you are succeeding, and the things this book asks you to protect, judgment and voice and learning and trust, are precisely the things that resist measurement and punish clumsy attempts at it. The playbook tells you where to aim; it does not tell you whether you are hitting anything, and the instruments that would tell you are the most treacherous in the whole framework, because they corrupt what they touch. So the next chapter takes on the hardest instrument problem in the book: how to measure the second ledger without destroying the very thing you are trying to measure.

### THE EVIDENCE

The framework travels everywhere because the four tensions arise from universal facts of human autonomy, safety, learning, and trust. The specifics, which tension dominates, the corrosive cadence threshold, the load-bearing skill, the available coaches, and the power distance, are local and must be discovered on site, never imported.

Dominant tensions cluster by industry: logistics carries specificity-autonomy and cadence-safety; healthcare carries cadence-safety and transparency-trust; software carries efficiency-learning. The method for any field is to identify the dominant tension, the load-bearing skill, the coaches available, and the power distance, then lead with the lever that answers the dominant tension.

### MONDAY MORNING

Write down four answers for your own organization: which of the four tensions is doing the most damage, which single skill your operation cannot survive without on its worst day, who could credibly coach without being an evaluator, and how steeply your hierarchy runs. Those four answers are your playbook in miniature. Lead with the lever that answers the first, and protect the skill named in the second before anything else.

CHAPTER FOURTEEN

# Measuring the Second Ledger

*What you do not measure, you will spend. But measure it the wrong way, and you will destroy the very thing you set out to protect.*

A leader I admired tried to measure the second ledger, and within a quarter he had destroyed the thing he was measuring.

He had read the argument, understood that judgment and voice and learning and trust were going unmeasured and therefore unprotected, and resolved to fix it the way a good operator fixes anything, by measuring it. He built a psychological-safety score. He surveyed it monthly, put it on a dashboard, set targets for it, and held managers accountable for it. And the score went up beautifully while the thing it was meant to track quietly died, because the moment safety became a number that managers were judged on, managers began managing the number. They coached their teams on how to answer the survey. They discouraged the kind of open dissent that might show up as a problem. The score rose, and the floor went quiet, and he had recreated, with his own measurement of the second ledger, the exact spotlight effect the second ledger exists to warn against.

This is the trap at the heart of this chapter, and it is a genuine paradox folded inside the larger one. The book has told you, repeatedly, that what you do not measure you will spend without noticing, which argues for measuring the second ledger. The book has also shown you, repeatedly, that measuring a human thing changes it, usually for the worse, which argues against. Both are true. The resolution is not to measure or to refuse, but to measure in a fundamentally different way than the systems do, and getting that difference right is one of the subtlest skills a leader of the algorithmic age has to learn.

## Why the Obvious Way Backfires

The obvious way to measure the second ledger is to do to it what the systems already do to the first, and that is exactly why it fails.

The systems measure the first ledger with high cadence, full coverage, individual attribution, and hard targets, and that combination is what makes them powerful and what makes them corrosive. Apply the same combination to trust or voice or learning and you get the corrosion without even the power, because these things, unlike throughput, are altered by the very act of being watched. A trust score that everyone is judged on becomes a performance of trust. A measured dissent rate becomes a managed dissent rate. A learning metric becomes a thing to optimize rather than a thing to do.

The second ledger is made of exactly the materials that curdle under measurement, which is why the first instinct, to instrument it like everything else, reliably destroys it.

This is Goodhart's old warning in a new place. When a measure becomes a target, it stops being a good measure, and the human qualities of the second ledger are the most Goodhart-vulnerable things in the entire organization, because they are subjective, self-reported, and easy to perform. So the rule for measuring the second ledger is the inverse of the rule for the first. Where the systems measure hard, you measure soft. Where they cover everything, you sample. Where they attribute to individuals, you look at patterns. Where they set targets, you refuse to. The second ledger is measured by doing nearly the opposite of what your dashboards do, which is why operators find it so counterintuitive and get it so wrong.

## The Measurement Paradox

It is worth stating the paradox of this chapter as plainly as possible, because it is a genuine contradiction and a leader has to hold both halves of it at once to navigate it.

Half one: what you do not measure, you will spend without noticing. This is the lesson of the whole book, the reason the second ledger matters at all. An organization that tracks only throughput will trade away judgment, voice, learning, and trust without ever seeing the trade on a report, because the trade has no line. By that logic you must measure the second ledger, or you will quietly liquidate it the way every metric-driven organization liquidates what its metrics cannot see.

Half two: measuring a human quality changes it, almost always for the worse. This is also the lesson of the whole book, the mechanism behind every tension in it. The act of measuring trust makes trust performable; the act of scoring safety makes safety a number to manage; the act of tracking voice turns voice into a managed metric. By this logic you must not measure the second ledger, or you will destroy it with the measuring, recreating in your own management the exact pathology the systems create.

Both halves are true, which is what makes this a paradox rather than a problem, and a paradox, as the whole book has insisted, is not solved but managed. You do not choose between measuring and not measuring. You measure in a way that escapes the trap, by inverting every property that makes the systems' measurement corrosive, and the rest of this chapter is the specific shape of that inversion. But hold the paradox first, because a leader who forgets half one stops paying attention to the second ledger and spends it blind, and a leader who forgets half two builds a trust dashboard and destroys what they meant to protect. The whole craft lives in the tension between the two, exactly as it does everywhere else in this book.

There is a quiet irony worth sitting with. The second ledger exists because the systems measure the first ledger too well, with too much cadence and coverage and consequence, and the harm comes from that very thoroughness. So when a leader sets out to measure the second ledger and reaches instinctively for thoroughness, more cadence, more coverage, more accountability, they are importing the disease into the cure. The instinct that makes someone good at managing the first ledger, the drive to measure it harder and better, is precisely the instinct that ruins the second. Measuring the second ledger well requires a leader to suppress their best-trained reflex, which is why it is so rarely done well, and why the leaders who manage the first ledger most impressively are often the ones who destroy the second most thoroughly when they finally turn their attention to it.

## Proxy, Do Not Pin

The first principle is to measure behaviors that indicate the thing, not the thing itself, and never to let the proxy become a target.

Do not measure trust. Measure whether people contest the system, because contesting is what people do when they trust that contesting is safe and useful, and gaming is what they do when they do not.

Do not measure psychological safety with a survey people learn to game. Watch the rate at which concerns and near misses get voiced, because voicing is the behavior safety produces. Do not measure learning with a test. Watch what happens when the system fails and someone has to step in, because the ability to step in is what learning leaves behind. The behaviors are harder to fake than the surveys, because they cost the person something and happen in the real flow of work rather than in a questionnaire designed to be answered.

And hold the proxies loosely. The instant you set a target for contests filed, people will file empty contests to hit it, and your proxy joins the first ledger as one more gamed number. The proxies are instruments for your own judgment, not goals for your people, and the difference must be real and visible. A contest rate that you watch to understand whether voice is alive is a healthy instrument. A contest rate that managers are measured on is a dead one within a quarter. The same number, used two different ways, is either a window or a weapon, and only one of them tells you the truth.

## Four Instruments for Four Ledgers

Each of the four unmeasured qualities has a behavioral proxy that resists faking better than any direct measure, and a leader can read all four lightly.

For judgment, look at decision quality rather than decision quantity, sampled through review of how a handful of real, hard calls were actually reasoned, not how many were made. A team whose judgment is intact reasons visibly and catches the system's errors; a team whose judgment has eroded defers to the system even when it is wrong, and you will hear that deference in how they explain their decisions when you ask. For voice, watch the rate at which dissent is not only raised but acted upon, because voice that is raised and ignored stops being raised, so a falling rate of acted-upon dissent is the early signal that the second ledger is emptying.

For learning, use the rotations themselves as the instrument: how do people perform when the tools are turned down, and is that performance holding or fading across cohorts. The rotation is already happening if you have built the third lever, so the measurement is free, and it is behavioral and unfakeable because the problem is either solved or it is not. For trust, watch the choice between contesting and gaming. People who trust a system improve it by flagging its errors; people who distrust it route around it and feed it what it wants. Which behavior dominates is the truest trust measure you will ever get, and it requires no survey at all, only attention to what people actually do with the system when they think no one is grading them on it.

## Measure Lightly, Measure Slowly

The cadence and weight of your second-ledger measurement must be the opposite of the systems', or it becomes a second surveillance apparatus and produces the same harm.

Measure infrequently. The first ledger refreshes every few minutes; the second should be read in quarters, because the things it tracks move slowly and watching them closely changes them. Measure by sampling rather than by total coverage, because you do not need to read every decision to know whether judgment is alive, and the attempt to read every one recreates the spotlight. Keep it qualitative where you can, because a conversation about how a hard call was reasoned tells you more about judgment than any score, and cannot be gamed the way a score can. The goal is a light, slow, sampled read that informs your judgment without ever becoming a dashboard your people have to perform for.

Above all, do not build a second-ledger dashboard. The temptation, once you have proxies, is to put them on a screen, refresh them, and manage them, which is to turn the cure into a second instance of the disease. The second ledger is not a dashboard. It is a discipline of attention, a set of questions a leader asks and answers with judgment, kept deliberately off the automated, high-cadence, target-driven machinery that does such violence to exactly these qualities. If you find yourself building a real-time trust tracker, stop, because you have begun measuring the second ledger the way the systems measure the first, and you already know how that story ends.

## The Instruments at Work

It helps to see each of the four instruments in use, because in practice they are quieter and more ordinary than the word measurement suggests.

Grace read judgment by sitting in occasionally on the way a handful of hard cases had been reasoned, not scored, just discussed: what did you see, what did the system say, how did you weigh them. A ward whose judgment was intact produced rich, specific reasoning and the occasional correct override of the system. A ward whose judgment had thinned produced deference, the same answer every time, the system said so. She needed only a few of these conversations a quarter to know which wards were keeping their judgment and which were quietly surrendering it, and no number could have told her what the reasoning revealed.

Marcus read voice by tracking not how many concerns were raised but how many were acted upon, because raised-and-ignored concerns stop being raised, so the acted-upon rate is the one that predicts whether voice is alive next quarter. Priya read learning straight off the rotations she already ran: when the tools came down, could people still do the work, and was that holding across cohorts or fading. And all three read trust by watching the choice their people made when the system was wrong, between flagging it and gaming it, because that single behavioral fork tells you more about trust than any survey, and it cannot be performed, because it happens in the real flow of work when no one thinks they are being graded on it.

## Two Ledgers, Two Leaders

The difference between measuring the second ledger well and badly is best seen in two leaders who set out to do the same thing.

The first, the one from the opening of this chapter, built a psychological-safety score, surveyed it monthly, set targets, and held managers accountable for it, and within a quarter the score was excellent and the safety was gone, managed into a performance the way any measured human quality gets managed into a performance. He had measured the second ledger exactly as the systems measure the first, with high cadence and hard targets and individual accountability, and he had gotten the predictable result, which was the corruption of the thing he measured.

The second leader wanted the same information and went the opposite way. She never built a score. Once a quarter she sampled a few teams, sat in on how some hard decisions had been reasoned, watched whether concerns that got raised got acted on, and noticed whether people were flagging the system's errors or quietly routing around them. She set no targets on any of it and put none of it on a dashboard. She simply read it, with her own judgment, and let it inform where she spent her attention. A year later she knew the true state of her second ledger and it was healthy, precisely because she had never subjected it to the machinery that would have falsified it. The same intention, two opposite methods, two opposite results, and the whole difference was whether the measurement borrowed the systems' logic or refused it.

## Sampling Without Surveilling

The single technical skill that separates a healthy second-ledger reading from a second surveillance apparatus is sampling, and most operators have to unlearn their instinct for coverage to do it.

The systems measure the first ledger by total coverage: every task, every worker, every minute, with nothing missed. That completeness is what makes them powerful and what makes the spotlight inescapable, because total coverage means there is nowhere a person is not being watched. The second ledger must be read the opposite way, by sampling, because you do not need to read every decision to know whether judgment is alive in a team, and the attempt to read every one rebuilds the very spotlight that kills the thing you came to measure. A few hard decisions, reasoned through in conversation, tell you the state of a team's judgment. You do not need all of them, and trying to get all of them does harm.

Sampling done well is light, rotating, and unpredictable in a way that reassures rather than threatens. You look at a few teams this quarter and different ones next, at a handful of decisions rather than a docket, and you make clear by your manner and your cadence that this is a leader trying to understand the health of the place, not an inspector building a file. The difference is felt immediately by the people involved, who can tell the difference between a leader sampling to learn and a system surveilling to score, and who respond to the first with candor and to the second with performance. The sampling is not a watered-down version of full coverage. It is a categorically different activity, aimed at understanding rather than control, and it has to feel that way to work.

The discipline that makes sampling safe is the refusal to aggregate it into a score. The moment you take your samples and roll them into a number, even for your own dashboard, you have started down the road to coverage, because a number begs to be made complete and comparable and tracked over time, and completeness and comparability and tracking are surveillance. Keep the samples as what they are, a handful of concrete observations held in judgment, and resist the pull to systematize them, because the systematizing is exactly how the cure turns back into the disease. A leader who can sample lightly, rotate the gaze, and refuse to aggregate has the one technical skill that lets them read the second ledger for years without ever once falsifying it.

If this feels unrigorous to an operator trained on full coverage and clean numbers, sit with the discomfort, because it is the right discomfort. The second ledger is not less rigorous than the first; it is rigorous in a different way, the way a skilled physician reading a patient is rigorous without running every test, the way an experienced judge of people is rigorous without a scorecard. The rigor is in the quality of the judgment applied to well-chosen samples, not in the completeness of the coverage, and a leader who cannot accept that kind of rigor as real will keep reaching for coverage and keep destroying what they measure. Learning to trust sampled judgment over complete measurement is, in the end, the same lesson the whole book teaches about the second ledger: the most important things are precisely the ones that punish the instinct to measure them like everything else.

## An Annual Reading

For most organizations, the right cadence for a deliberate second-ledger reading is roughly annual, slow enough not to distort and regular enough to catch drift.

Once a year, set aside the dashboards and ask the four questions directly, with judgment rather than instruments. Is judgment alive: when I look at how hard decisions are actually reasoned, do I see thinking or deference. Is voice real: is the rate of acted-upon dissent holding, and can people point to things that changed because they spoke. Is learning maintained: when the tools come down, can people still do the work, across every cohort and not just the veterans. Is trust intact: when the system is wrong, do people flag it or game it. Four questions, answered from sampled behavior and honest conversation, once a year, written down where your decisions can reach them.

Keep this reading deliberately separate from the performance and operating reviews, because the moment it touches evaluation it corrupts, and the whole point is to have one account in the organization that is not subject to the logic of the dashboard. The annual second-ledger reading is not a scorecard and produces no ranking. It is a leader sitting with the hardest questions about the health of the human capacities the systems erode, refusing the false comfort of a number, and accepting the harder work of knowing through judgment. That refusal is the discipline, and it is the only way to measure the second ledger without destroying it.

## Who Holds the Measure

The final principle is about custody. The second ledger is held by the leader, never handed to the system.

Everything that makes the systems good at measuring the first ledger, the automation, the consistency, the tirelessness, makes them dangerous custodians of the second, because the second ledger requires exactly the contextual, forgiving, pattern-seeing judgment the systems do not have. So you keep it. You read the proxies, you have the conversations, you sample the decisions, and you integrate it all with judgment that no dashboard could replicate, and you use it to inform your decisions rather than to automate them. The moment you automate the second ledger, you have lost it, because you have asked the thing that cannot see judgment to keep the account of judgment.

Notice what the real instrument turns out to be in all of this. It is not a survey, a score, or a dashboard. It is the leader's own judgment, applied to well-chosen samples of real behavior, refusing the false comfort of a number. The second ledger is read by a person paying disciplined attention, which is both its great limitation and its only protection, because attention can see what a metric cannot and cannot be gamed the way a metric is. A leader who wants to measure the second ledger is really being asked to develop and trust their own judgment about the human health of their organization, which is harder than building a dashboard and is the only thing that works.

Guard, finally, against over-reading. The opposite error to ignoring the second ledger is becoming obsessed with it, turning every conversation into an assessment and every observation into evidence, until the reading itself becomes a low-grade surveillance that your people can feel. The second ledger is read lightly, in passing, a few times a year, and then mostly left alone, because the things it tracks are healthiest when they are trusted rather than constantly examined. A leader who reads it once a quarter and acts on what they learn has it right; a leader who reads it constantly has turned a discipline of attention into a new anxiety, and the anxiety leaks, and the leaking does its own quiet harm to the very qualities the reading was meant to protect. The right posture is closer to that of a gardener who checks the soil now and then than that of a manager who watches a screen, trusting the thing to grow when it is tended and left alone, and intervening only when the occasional reading shows it is needed.

The principle to carry out of this chapter is the most delicate in the book. Measure the second ledger, because what you do not measure you will spend, but measure it by its opposite: softly, slowly, by sampled behavior rather than gamed survey, held in your own judgment rather than in a machine, and never, ever as a target. The second ledger is the one account you keep precisely so that not everything in your organization is subject to the logic of the dashboard, and the fastest way to lose it is to subject it to that logic too.

Measuring the second ledger tells you whether the levers are working. But knowing they can fail raises a harder question than measurement, which is what failure actually looks like up close, because the levers have bad twins that resemble them from the front, and a leader who cannot tell the real thing from its counterfeit will build the counterfeit and wonder why it poisons the well. And the counterfeit is not the rare exception; it is the common case, the default a busy leader drifts into without ever deciding to. So the next chapter is about everything that goes wrong, the anti-patterns, the failure modes, and the specific betrayals that turn these practices from a cure into the most corrosive thing of all.

### THE EVIDENCE

The second ledger must be measured by the inverse of the systems' logic. Where dashboards measure with high cadence, full coverage, individual attribution, and hard targets, the second ledger is read slowly, by sampling, by pattern, and never as a target, because judgment, voice, learning, and trust curdle under the measurement that throughput tolerates. This is Goodhart's warning applied to the most Goodhart-vulnerable qualities in the organization.

Use behavioral proxies that resist faking: contesting versus gaming for trust, the rate of acted-upon dissent for voice, performance with the tools turned down for learning, and sampled review of how hard calls were reasoned for judgment. Hold proxies loosely as instruments for the leader's judgment, never as targets, and keep them off any automated dashboard.

### MONDAY MORNING

Pick one quality from the second ledger and one behavioral proxy for it: trust and the contest-versus-game choice is a good place to start. Spend a quarter simply watching that behavior, with no target attached and no dashboard built, and let it inform your judgment about whether voice is alive. If you feel the urge to set a number on it or refresh it weekly, that urge is the disease, and resisting it is the discipline.

CHAPTER FIFTEEN

# When It Goes Wrong

*Every good idea has a bad twin that looks just like it from the front. The levers are no exception, and the twin is more dangerous than no idea at all.*

The most damaging thing I ever watched a leader do was run a calibration session perfectly and mean none of it.

He had read about the lever. He gathered the team, put the metric on the screen, asked the three questions, and listened, or appeared to. People spoke. They told him, honestly and at some risk, what the metric punished and what it missed. He thanked them warmly. And then nothing changed, because he had already decided what the metric would be before he walked in, and the session had been a way to make the decision feel participatory rather than a way to actually improve it. The team understood this within about a week, when the unchanged metric appeared exactly as it had always been, and what he had built was not calibration. It was the appearance of calibration laid over an unchanged reality, and it taught his people that voice in his organization was a performance they were expected to participate in and then ignore.

That is worse than never having asked, and it is the theme of this chapter. Each of the five levers has a counterfeit that looks identical from the front and does the opposite underneath, and the counterfeits are more dangerous than inaction, because they spend the credibility that the real levers run on. A leader who does nothing leaves the well full. A leader who builds the counterfeit poisons it, and a poisoned well is far harder to recover than an empty one. So it is worth knowing the bad twins by sight.

## The Bad Twin

The reason these failures are so dangerous is that the counterfeit and the real thing are indistinguishable from the front, and you only ever see the front while you are building it.

A real calibration session and a fake one look identical in the room. The same people, the same metric on the screen, the same three questions, the same warm thanks at the end. The difference is entirely in what happens afterward and in what was in the leader's head beforehand, neither of which is visible at the moment of the meeting. This is what makes the counterfeit so seductive and so common. You cannot tell, while you are running it, whether you are building the real thing or its bad twin, because the only evidence that distinguishes them, the actual change, the genuine willingness to be corrected, lies in the future or in your own intentions, and both are easy to misjudge in your own favor.

Leaders therefore deceive themselves about which one they are building, routinely and sincerely. The leader running the empty calibration session believed he was running a real one, right up until the moment of the inconvenient change, when he chose the comfortable status quo and revealed, to his people and eventually to himself, which twin he had built. The intent that distinguishes the real lever from its counterfeit is not always known even to the leader until it is tested by a concrete choice between the loop and their own convenience. Which means the only honest way to know which you are building is to watch what you do at the decision points, not what you intend at the outset, because the intent is exactly the thing you cannot trust about yourself here.

This also explains why the counterfeits propagate so easily as best practice. Since the real thing and the bad twin present identically, an organization can showcase its counterfeit in perfect good faith, describe it accurately as far as the description goes, and have other leaders copy the visible form, which is all that was ever transferable anyway. The form copies easily; the substance does not copy at all, because the substance is a standing willingness to pay a price, and willingness cannot be packaged in a case study. So the counterfeit spreads faster than the real thing, the way the cheaper version of anything outcompetes the dearer, and the field fills with hollow practices that everyone can see the front of and no one can see behind. Knowing this, you evaluate every practice, including your own, by its back rather than its front, by the trail of real changes behind it rather than the impression it makes in the moment.

The deepest version of the warning is this: the counterfeit is not a different practice from the real thing. It is the same practice with the substance quietly removed, which means you can convert a real lever into a counterfeit without changing anything visible, simply by starting to fold on the hard calls while keeping the meetings on the calendar. A loop does not announce its own corruption. It keeps its form while losing its function, and the leader, seeing the unchanged form, can believe nothing has changed long after everything has. The vigilance this demands is permanent, because the drift toward the counterfeit is permanent, powered by the constant pressure to keep the cheap form and shed the expensive substance, and the only defense is to keep checking the back of your own practices for the trail of real changes that is the only thing that was ever real about them.

## The Five Counterfeits

Each lever fails in a characteristic way, and the failure always comes from keeping the form while abandoning the substance.

Calibration becomes theater, the session that gathers input and changes nothing, as above.

Coaching becomes surveillance, when the person coaching you turns out to be feeding your evaluation, so the safe space was a trap and the questions were a way to build a file, which teaches people to perform for the coach exactly as they perform for the system. Rotation becomes busywork, when the practice is hours logged on problems that are hard in appearance only, a simulator everyone knows is theater, which builds no skill and breeds the cynicism that any fake exercise breeds. Each of these keeps the visible shape of the lever and hollows out the thing that made it work.

Explainable dashboards become data dumps, when transparency is offered as a flood of logic and weights with no Action step and no recourse, which produces Lena's glass box, comprehension without trust, and is sometimes worse than the black box it replaced. And voice forums become grievance sessions, when the forum has no real scope to change anything and so curdles into a place people come to complain and leave more bitter, or worse, becomes a stage where the loud dominate and the system gets captured by a faction. In every case the diagnosis is the same: the leader kept the ritual and discarded the point, usually because the ritual is cheap and the point is expensive, and the expense was the whole reason the lever worked.

## The Three Meta-Failures

Above the individual counterfeits sit three failures that can kill the whole loop regardless of how well any single lever is built.

The first is doing it for show, building the practices to look like a modern, enlightened organization rather than to actually change anything, which your people detect with depressing speed because the tell is always the same: nothing closes. The second is doing it once, the brilliant launch that is never sustained, which converts a relationship back into a one-time survey and spends the trust it briefly built, the failure the ninety-day chapter warned against. The third is the most corrosive, and it deserves its own treatment, because it is not mere neglect. It is the deliberate use of the levers as instruments of control rather than of voice.

These three are deadly because they operate at the level of intent, and intent leaks. You cannot fake meaning it over time, because the moment a real decision has to be made, between the inconvenient change the loop calls for and the comfortable status quo, your true intent shows, and your people read it instantly. The mechanics of the levers can be taught. The intent to actually share power, to actually be corrected by what you hear, cannot be faked for long, and the levers built without it become elaborate machines for teaching people that their voice is decorative.

## When the Levers Are Weaponized

The darkest failure mode is not neglect but weaponization, the use of a lever to do the opposite of what it promises, and it poisons the well in a way that ordinary failure does not.

A voice forum used to identify the dissenters, so that speaking up marks you for management rather than empowering you. A coaching relationship used to gather evidence for a termination already decided. A calibration session used to launder a metric the leader had no intention of changing, manufacturing the consent of the governed after the fact. An explainability initiative used to demonstrate that the system is fair so that complaints can be dismissed as uninformed. Each of these takes the trust the lever asks people to extend and uses it against them, which is the one move that does not merely fail but actively betrays.

Weaponization is worse than every other failure because it is not recoverable on the same terms. A team that experienced theater can be won back by a leader who later means it, slowly. A team that was betrayed, whose openness was turned into a weapon against them, will not extend that openness again for years, possibly not to anyone, because they have learned that the safe space was bait. If you are not prepared to actually share the power these levers imply, do not build them, because the half-built version that gathers trust and then exploits it is the single most destructive thing in this book, more destructive than the unaccompanied system you started with, because at least the system never pretended to be on their side.

## Why the Counterfeit Is So Tempting

No one sets out to build a counterfeit. Leaders drift into them, and understanding the drift is how you resist it, because the pull is real and constant.

The counterfeit is tempting because it is cheaper, faster, and indistinguishable from the real thing at the moment you build it. The form of a lever, the meeting, the pairing, the dashboard, the forum, costs almost nothing and can be stood up in a week. The substance, actually changing the metric, actually protecting the time, actually being corrected by what you hear, costs real power and real inconvenience and never stops costing them. So at every decision point the counterfeit is the path of least resistance, and a leader under pressure, who genuinely means well, can assemble an entire counterfeit loop one reasonable shortcut at a time without ever deciding to fake anything.

This is why good intentions are no protection. The leader who ran the empty calibration session was not a cynic; he was a busy person who had already decided the metric and did not want to reopen it, and gathering input felt like a courtesy rather than a betrayal. The drift into counterfeit is paved with small, defensible choices, each of which keeps the form and trims a little substance, until one day the practice is hollow and the leader is genuinely surprised that no one believes in it. The defense is to watch the substance specifically, at every step, and to notice when you are keeping a ritual while quietly declining to pay its price, because that noticing is the only thing that interrupts the drift.

## The Slow Poison and the Fast

Counterfeits poison the well at two speeds, and the fast poison is in some ways less dangerous than the slow, because at least you can see it work.

The fast poison is the obvious betrayal: the forum that is openly ignored, the coaching that is revealed to feed evaluation, the calibration that visibly changes nothing. People react quickly and the damage is sharp, but it is at least legible, and a leader paying attention can see the trust collapse and know they caused it. The slow poison is worse precisely because it is invisible. It is the practice that mostly works, that closes some loops and quietly drops others, that means it on the easy questions and folds on the hard ones. It does not provoke a collapse. It produces a slow leak, a gradual return to the quiet, a creeping cynicism that the leader attributes to anything but the cause, because no single broken promise was large enough to notice.

The slow poison is the more common death of a once-real loop, and it is the reason the discipline of closing every loop, every time, matters so much. A loop that closes ninety percent of the time is not ninety percent as good as one that closes always; it is teaching, in the ten percent, that voice here is conditional, and conditional voice is the kind people stop spending. The standard has to be every time, not usually, because the slow poison enters through the gap between usually and always, and it enters so gradually that by the time the leader notices the floor has gone quiet again, they have often forgotten that they once promised it would not.

## Spotting Counterfeits in the Wild

You will encounter these counterfeits in other organizations, sometimes praised as models, and learning to spot them protects you from importing a hollow practice dressed as a success.

When an organization boasts of its listening sessions, its coaching culture, its transparency, ask the one question that cuts through every counterfeit: what changed because of it, specifically, and can the people who spoke point to the change. A real practice produces a stream of specific, attributable changes that the people involved can name. A counterfeit produces testimonials about how valued people feel and no specific changes anyone can point to, because there were none. The tell is always the same, the gap between the warmth of the description and the absence of concrete consequence, and once you can see it you will see it everywhere, including in places that have won awards for it.

This matters because counterfeits are contagious through imitation. A celebrated counterfeit becomes a model that other leaders copy, importing the form that was all there ever was, and the failure propagates dressed as best practice. The protection is to evaluate any practice, your own or another's, by its consequences rather than its description, by what visibly changed rather than by how good it sounded in the case study. The real thing and the counterfeit give identical presentations. They differ only in the trail of actual changes they leave behind, and that trail, or its absence, is the only evidence that cannot be faked.

## How to Tell If Yours Is Real

There are a few honest tests a leader can apply to their own practices, and they are uncomfortable on purpose.

Does anything actually change because of it, and can your people point to what. Can the people who spoke see the effect of their speaking, or does the input vanish into a process they never hear from again. Do the quiet people speak, or only the safe and articulate ones, because a practice that only the comfortable use is not reaching the truths that matter. And the hardest test of all: would you run it if you knew it would produce bad news about you, specifically. A leader who would quietly prefer the forum not surface their own mistakes has already built a counterfeit, whatever its mechanics, because the willingness to be corrected by what you hear is the substance, and everything else is form.

Apply these tests to yourself before your people apply them for you, because they will, and they are better at it than you are, having far more at stake in the answer. The tests are uncomfortable because the real thing is uncomfortable. A genuine loop regularly produces inconvenient changes, bad news about the leader, and decisions the leader would not have made alone, and a leader who finds their own loop suspiciously comfortable should suspect they have built the counterfeit, because comfort is the smell of a practice that has stopped telling the truth.

## Counterfeits You Inherit

Sometimes the counterfeit is not one you built but one you walked into, a hollow practice already running when you arrived, and that situation has its own hazards.

Many leaders inherit a forum that meets and changes nothing, a coaching program that everyone knows feeds evaluation, a calibration ritual that launders decisions already made. The practices exist, they are on the calendar, they have names and owners, and they are counterfeits, and your people know it even if no one says so. The first hazard is that you mistake the existing form for a foundation to build on, when it is in fact a liability to clear, because a counterfeit running under your predecessor has already taught your people the lesson you now have to overcome, and the practice carries that lesson forward every time it meets and changes nothing.

The instinct on arriving is to improve the inherited practice, to make the forum a little better, the coaching a little safer, and that instinct is usually wrong, because incremental improvement of a counterfeit keeps the poisoned name and the poisoned expectations attached to it. It is often better to end the counterfeit cleanly and visibly, to say plainly that the old forum did not work and is over, and then to start something new and small that you intend to keep real, than to try to rehabilitate a practice your people have already learned to discount. The name is contaminated; the expectations are set; and a fresh, modest, real loop with a new name will outrun a refurbished counterfeit every time, because it is not carrying the old lesson on its back.

Whatever you do, do not let an inherited counterfeit keep running unexamined simply because it predates you and ending it would be awkward. Every meeting it holds that changes nothing is a fresh withdrawal from the account, made in your name now whether you built the practice or not, because your people no longer distinguish between the predecessor's empty forum and your tolerance of it. Inheriting a counterfeit is inheriting a debt, and the debt keeps accruing until you either make the practice real or end it. Pretending you did not notice, because you did not create it, is not neutral; it is the slow poison continuing under new management, and your people will read your tolerance of it as your endorsement of it. The first real act of leadership in a place full of inherited counterfeits is often simply to name one of them honestly, out loud, as the empty thing it has become. That naming costs you nothing but the small discomfort of admitting the obvious, and it tells your people something they have been waiting to learn, which is that the new leader can at least see what they have all long since seen, and is not going to pretend otherwise.

## Recovering From a Poisoned Well

If you have already built a counterfeit, knowingly or not, the well is poisoned, and recovering it is slower and humbler than building it would have been if you had done it right the first time.

You cannot announce your way out of a poisoned well, because announcement is exactly what your people have learned to discount. You recover the way you should have started: smaller, slower, and with proof. Name the failure honestly, without the defensiveness that will confirm their worst read of you, because people forgive an owned mistake far more readily than a denied one. Then pick one small thing, close one real loop, and let the evidence do the work that no words can, over a span of time that will feel insultingly long to you and merely appropriate to them. Trust spent is recovered at a fraction of the rate it was lost, and the only currency that buys it back is kept promises, accumulated slowly, with no shortcut available. The leader who poisoned the well, or inherited a poisoned one, has to become, for a long while, almost boringly reliable, promising little and delivering all of it, until the slow arithmetic of kept promises finally outweighs the broken ones in their people's memory, which takes far longer than anyone wants and is the only thing that works.

What all the counterfeits finally have in common is that they require something the real thing does not: a leader willing to keep paying a price that never stops coming due. The mechanics of every lever can be learned in an afternoon. The willingness to be corrected by what you hear, to change the comfortable thing because someone with less power than you said you should, to keep doing it in the hard quarter when no one would blame you for quietly letting it lapse, cannot be learned at all. It can only be chosen, repeatedly, and that repeated choosing is the entire difference between the real thing and its twin.

This is why so many leaders who understand this book perfectly still build counterfeits. Understanding is not the scarce resource. The scarce resource is the standing willingness to give up power and comfort, again and again, with no end date, in service of something most of the organization cannot see and the dashboards will never credit. A leader who has the knowledge and not the willingness will build immaculate counterfeits, well-designed and hollow, and will be sincerely puzzled at why they do not work. The knowledge in this book is necessary and nowhere near sufficient, and the gap between the two is filled only by character, which is the one thing no book can supply.

The principle to carry out of the failures is that the form of these levers is cheap and the substance is expensive, and your people are paying attention only to the substance. The counterfeit costs little to build and poisons the well; the real thing costs real power and fills it. Before you build any lever, ask whether you are prepared to pay its real price, which is to actually be changed by what you hear, because if you are not, the honest move is to leave the well full rather than poison it with a practice you do not mean.

Knowing how the levers fail protects you from building the counterfeits. But there is one more place the whole enterprise can die, and it is not on the floor where the work happens. It is in the boardroom, where directors and executives who see only the first ledger will look at the cost of the loop and the slowness of the rotations and ask, reasonably, why you are spending money to make the organization slower. It is a fair question, asked in good faith, by people doing their jobs, and it can undo in a single meeting everything the floor took years to build. Winning that argument is its own skill, distinct from everything that came before, and the final chapter of this part is about how to have the board conversation without losing the loop to the very people who govern it.

### THE EVIDENCE

Each lever has a counterfeit that keeps the form and discards the substance: calibration as theater, coaching as surveillance, rotation as busywork, dashboards as data dumps, forums as grievance or capture. Counterfeits are worse than inaction because they spend the credibility the real levers run on, poisoning a well that doing nothing would have left full.

Three meta-failures kill the loop regardless of mechanics: doing it for show, doing it once, and weaponizing it. Weaponization, using a lever to identify dissenters or launder a decision, betrays trust unrecoverably. The honest tests are whether anything changes, whether people can see it change, whether the quiet speak, and whether the leader would run it knowing it would produce bad news about themselves.

**MONDAY MORNING**

Apply the hardest test to one practice you already run: would you keep running it if you knew, for certain, that it would surface a serious mistake of your own. If the honest answer is no, you have built a counterfeit, and your people likely already know. Either commit to being corrected by what it surfaces, starting with the next inconvenient thing it tells you, or shut it down honestly rather than let it keep gathering a trust you do not intend to honor.

CHAPTER SIXTEEN

# The Board Conversation

*The hardest person to convince is not the worker on the floor. It is the director in the boardroom, who only ever sees the first ledger.*

Priya nearly lost the entire loop in a forty-minute budget review, to a question she had no good answer ready for.

The board was supportive, in the way boards are supportive of things that do not cost anything. Then the rotation program came up, the one week in six that engineers spent on hard problems with the tools turned down, and a director did the arithmetic out loud. That is roughly seventeen percent of your most expensive people's time, he said, spent not shipping. In a competitive market. Can you justify that. And Priya, who could have justified it completely, found herself reaching for the language of the second ledger, judgment and resilience and developmental space, in a room that ran entirely on the first, and watched the words land as soft and unconvincing exactly where she most needed them to be hard.

She kept the program, barely, and only by promising numbers she had not yet learned how to produce. But the episode taught her the lesson of this chapter, which is that the loop has one more constituency beyond the workers and the managers, and it is the one with the power to cut the whole thing in a single meeting. The board, the executive committee, the people who govern by metric and have never set foot on the floor, see only the first ledger by the very nature of their role, and a leader who cannot translate the second ledger into the language of the first will lose the loop to the people who are supposed to be protecting the organization's long-term health.

## The Highest-Leverage Conversation

Of all the conversations in this book, the board conversation is the one with the most leverage and the most asymmetry, and understanding that asymmetry is what makes a leader take it as seriously as it deserves.

Everything else in this book is built from below, slowly, one closed loop at a time, over quarters and years. The board conversation can undo all of it in forty minutes. That is the asymmetry: the loop is constructed gradually and can be destroyed instantly, defended daily from below and killed in a single meeting from above. A leader who has spent two years building credibility with their people, closing loops, protecting rotations, earning the trust that makes the whole system run, can watch it all be defunded in one budget review by people who never saw any of it and meant no harm, because to them it was a line item with a soft justification and a hard cost. The work of years is hostage to a conversation of minutes, and that should focus the mind.

This asymmetry is not a flaw in your board. It is the structure of governance, and it would be true of any competent oversight body. Boards exist to allocate capital and manage risk at the highest level, which means they exist to cut things that do not justify their cost, which means a practice that cannot justify its cost in the board's language is exactly the kind of thing they are supposed to cut. The board doing its job is the threat. So the leader cannot treat the board conversation as an afterthought to the real work of building the loop. It is part of the real work, the part that protects everything else, and a leader who is brilliant at building loops and hopeless in the boardroom will lose their loops to their own incompetence at the one conversation that governs them all.

Priya learned this the hard way, in the forty minutes that nearly cost her the rotation program. She had treated the board as a friendly audience for work she assumed they valued, and discovered, too late in the meeting to recover gracefully, that they valued it only as long as it cost nothing, and that the moment it cost something visible she had no language ready to defend it. The lesson was not that the board was hostile. It was that she had failed to prepare for the one conversation that could undo two years of work, because she had not taken it as seriously as she took the work itself, and the two deserve equal seriousness precisely because one can erase the other.

There is a second, subtler stake. The board conversation is not only about protecting your own loop. It is about whether the second ledger gets governed at all, anywhere in the organization, because the board sets the terms by which everything below it is judged. A board that governs purely by the first ledger pushes first-ledger thinking down through every layer beneath it, so that every manager learns to optimize what the board can see and spend what it cannot. Changing what the board pays attention to, even a little, changes the incentives of the entire organization beneath it. The board conversation is therefore the highest-leverage point in the whole system, the one place where a small change in what gets attention propagates downward through every level, which is why it is worth far more preparation than its forty minutes would suggest.

All of which means the board conversation is not a defensive chore to be survived but an offensive opportunity to be used. Handled badly, it is where the loop dies. Handled well, it is where the second ledger gets its only foothold at the level that governs everything else, where the organization's definition of what counts can be widened at the source. The leader who walks in prepared is not merely protecting a budget line. They are trying to change, at the one point where it propagates furthest, what the organization considers worth seeing. The rest of this chapter is about how to walk in that prepared, starting with why the board cannot see the second ledger on its own.

## Why the Board Sees Only the First Ledger

This is not a failing of bad boards. It is structural, and treating it as a failing rather than a structure is the first mistake.

A board governs an organization it cannot see directly, through instruments, which means through metrics, which means through the first ledger almost exclusively. The second ledger, judgment and voice and learning and trust, is invisible at the altitude a board operates at, not because directors are foolish but because the information that reaches them has already been compressed into the numbers that survive the trip upward. By the time the state of the organization reaches the boardroom, it is a dashboard, and the dashboard is the first ledger by construction. The board is, in a sense, the ultimate victim of the competence paradox, governing entirely by the measured half of a reality whose unmeasured half is exactly where the risk lives.

Understanding this changes how you make the case. You are not trying to convince the board to care about soft things, which will fail, because their entire apparatus is built to discount what has no number. You are trying to translate the hard reality of the second ledger into the language the board already speaks, which is the language of risk, cost, and competitive position. The second ledger is not soft. It is hard reality that happens to lack a number, and your job in the boardroom is to supply the hardness the number would have supplied, in terms the board cannot dismiss as sentiment.

## Translate Everything Into Risk

The language that lands in a boardroom is the language of risk, and almost everything in this book translates into it cleanly once you try.

The rotation program is not lost velocity. It is insurance against the low-probability, high-cost event of a critical system failing with no one able to fix it, the outage that threatens the revenue line, priced at one week in six of a few engineers against the cost of the failure it prevents. The coaching layer is not a soft benefit. It is mitigation of the catastrophic error, the bed-nine event, the safety incident, whose expected cost dwarfs the program many times over. The calibration room and the voice forums are not participation theater. They are early-warning systems for the metric corrosion and the quiet attrition that, left undetected, become the mass departure of your best people and the slow rot of your core asset. Every lever is a risk control, and boards understand risk control, because managing risk is most of what a board is for.

The reframe is total once you commit to it. Stop describing the levers as ways to treat people well, which is true and unpersuasive in that room, and start describing them as the management of specific, expensive, delayed risks that the organization is currently carrying uninsured. The board is not being asked to fund a values program. It is being asked to stop running a critical operation with no hedge against its most predictable failures. That is a sentence a board can act on, and it is the same loop described in the only dialect the room speaks fluently.

## The Numbers You Can Actually Show

You will be asked for numbers, and the honest move is to show the ones you have rather than invent the ones you do not.

You have the leading indicators from the launch, the costly participation that signals the loop is alive, and you can show them trending. You have the cost of the failures the levers prevent, which can often be estimated from your own history, Carmen's turnover cost and the cascade behind it, the liability exposure of a near miss, the revenue at risk in an unfixable outage. And you have the cost of the levers themselves, which is almost always small against the failures, and the contrast between the two is the entire argument. You do not need a clean return-on-investment figure for trust. You need to put the cost of the hedge next to the cost of the risk and let the board see that the hedge is cheap.

Be honest about what you cannot measure, because a board will trust you more for the candor and because the inability to measure something is not evidence of its absence. Tell them plainly that the second ledger resists clean measurement, that you read it by the proxies and the behaviors described earlier, and that the absence of a precise number is exactly the condition that lets these risks accumulate unpriced. The leader who admits the measurement is imperfect and shows the proxies anyway is more credible than the one who arrives with a suspiciously precise figure for the value of psychological safety, which any competent director will rightly distrust.

## The Three Objections and Their Answers

Three objections come up in every boardroom, and a leader who has the answers ready does not get ambushed the way Priya did.

The first is that it slows us down. The answer is that it slows the easy half of the work, the shipping and the throughput, in order to protect the hard half, the judgment and resilience, and that competitors who coast on the easy half will look faster right up until the day the hard half is needed and they discover they spent it. Speed that is borrowed from the reserve is not speed. It is debt, and the bill comes due at the worst possible moment, which is precisely when the system fails and there is no one left who can fix it.

The second objection is that it is not measurable, and therefore not manageable. The answer is that the risk you are already carrying is not measurable either, and you are managing it anyway, by ignoring it. The absence of a number is not the absence of a cost; it is the absence of visibility into a cost that is accruing regardless. The third objection is the seductive one: can we not simply buy a better system that does not have these problems. The answer is that the better system has the same paradox, only sharper, because a more capable system optimizes the person harder and erodes the second ledger faster. There is no version of the technology that escapes the paradox, because the paradox is in the relationship between any optimizing system and a human being, not in the quality of the system. Buying a better one buys you a steeper version of the same problem.

## The CFO Is Not the Board

It helps to know which first-ledger audience you are actually facing, because the chief financial officer, the chief executive, and the board each see the second ledger from a different blind spot.

The chief financial officer thinks in cost and efficiency and will press hardest on the velocity objection, the seventeen percent of expensive people not shipping, so the case that lands is the one that prices the hedge against the specific failure it prevents, in the currency of dollars at risk. The chief executive thinks in competitive position and strategy and is most moved by the argument that competitors coasting on the easy half will look faster until the day they are not, which is a strategic risk rather than a line item. The board thinks in governance and fiduciary duty and responds to the framing that the organization is currently carrying a material, predictable risk with no hedge, which is precisely the kind of thing a board exists to notice and is failing to, because the risk has no number on the dashboard they govern by.

Tailor the same underlying case to the seat. The loop has not changed; the dialect has. To the chief financial officer, it is risk-adjusted cost management. To the chief executive, it is durable competitive advantage that the coasting competitor is quietly forgoing. To the board, it is responsible governance of a risk the current instruments cannot see. A leader who brings the same generic pitch to all three will land it with none, and a leader who knows which blind spot they are addressing can supply exactly the hardness that seat is missing. The art is not three different arguments but one argument translated three ways, into the native concern of whoever holds the power to cut the loop.

## Pricing the Unpriced Risk

You will be pushed to quantify, and the honest, effective move is to price the failure rather than the virtue, because the failure has a number even when the virtue does not.

You cannot put a credible figure on the value of trust, and you should not try, because any number you invent will be rightly discounted. But you can usually price the failures the levers prevent, using your own history and ordinary risk arithmetic. What did the last unfixable outage cost, or what would one cost, in revenue and recovery and reputation, and how likely is it given a thinning senior bench. What is the liability and human cost of one serious safety event of the kind the coaching layer prevents. What did losing your last several best people actually cost, fully loaded, including the cascade behind each departure. These are estimable, and the estimate, set against the small known cost of the lever, is the entire argument: here is what we are exposed to, here is what the hedge costs, the hedge is cheap.

Be transparent about the uncertainty in the estimate, because a board that does this for a living will trust a range offered honestly over a point estimate offered confidently, and the honesty itself is persuasive. The goal is not a precise return-on-investment figure, which would be false precision and would invite an argument about the decimals. The goal is to move the conversation from is this worth it, where the unmeasured benefit loses to the measured cost, to what is our exposure and what does covering it cost, where the measured failure cost makes the small lever cost look obviously worth paying. You win by changing the question from the value of a virtue to the price of a risk, because the second question is one the first ledger can actually answer.

## What Not to Say

There is a vocabulary that loses the room reliably, and avoiding it is half the battle, because the wrong words trigger the discount before your argument is even heard.

Do not lead with the language of the second ledger in its own terms. The words judgment, voice, developmental space, psychological safety, and trust are true and necessary, and in a first-ledger room they land as soft, as the very thing the room is built to discount, and they prime your audience to file the whole proposal under nice-to-have. Priya lost her footing the moment she reached for them. Save that vocabulary for the conversation you have after you have won on risk, once the levers are funded and you are explaining what they protect. Open in the room's own language, which is risk and cost and competitive position, and you will be heard; open in yours, and you will be politely discounted before you finish the sentence.

And do not apologize for the cost or frame it as a sacrifice the organization makes for its people, which invites the question of whether it can afford the generosity right now, and the answer in a hard quarter is always no. Frame it as what it is, the prudent management of a risk the organization is currently running uninsured, which is not generosity but competence, and not optional but overdue. The leader who walks into the boardroom apologizing for the cost of the loop has lost before they start. The one who walks in slightly incredulous that the organization has been carrying this risk unhedged for so long has the right posture, because it is the accurate one.

## The Annual Risk Letter

The most durable way to keep the board's attention on the second ledger is to give them a regular, plain accounting of it in the language of risk, before they ask and before the budget pressure forces the question.

Once a year, write the board a short, honest accounting of the risks the organization is carrying that the operating metrics do not show: the state of the senior bench and what a thinning one exposes, the health of voice and what its erosion would cost, the resilience of the workforce when the systems fail. Frame each as a risk with an estimated exposure and a named hedge, the relevant lever, with its small cost set beside the large one it covers. This is not a values statement and should not read as one. It is a risk report, in the genre the board already consumes, covering the specific category of risk their other instruments are structurally blind to.

The letter does two things at once. It keeps the second ledger continuously visible at the level that governs everything, so that the loop is not rediscovered and re-litigated in every budget cycle as if it were a new indulgence, but understood as the standing management of a standing risk. And it slowly trains the board to expect this category of information, to ask after it, to treat the health of judgment and voice and learning and trust as part of what they govern rather than as a soft extra. A board that receives the risk letter every year, and comes to expect it, has been quietly converted into a custodian of the second ledger, which is the whole aim, accomplished not by a single persuasive meeting but by a steady habit of accounting.

Do this before you are forced to, because the difference between offering the accounting proactively and producing it under budget pressure is the difference between governing and defending. A risk letter delivered annually, in calm, establishes you as the leader who sees the risks the instruments miss and manages them responsibly, and it makes the eventual budget challenge far easier, because the board already understands what the lever hedges and why. A justification produced reactively, in the heat of a review, is Priya in her forty minutes, reaching for soft words in a hard room. The letter is how you make sure that conversation, when it comes, happens on ground you prepared rather than ground you are scrambling to hold.

## Make the Board an Ally

The goal is not to survive the board conversation but to convert the board into a custodian of the second ledger, so that the loop is protected from above rather than perpetually defended from below.

This means giving the board its own version of the second ledger, a small set of leading indicators reported upward alongside the operating metrics, so that the health of judgment, voice, learning, and trust becomes something the board governs rather than something it unknowingly undermines. It means planting one board-level question that reframes their oversight, the question Marcus learned to ask of himself, asked now at the top of the organization: what is our system failing to see, and what are we doing about it. A board that asks that question routinely has become an ally, because it has internalized the second ledger as part of its governing responsibility rather than treating it as a line item to cut.

The payoff of getting this conversation right compounds in a way the others do not. A board converted into a custodian of the second ledger does not merely spare your loop one budget cycle. It protects the loop for years, asks the right question of every leader beneath it, and slowly reshapes what the whole organization treats as worth seeing, because the board sets the terms that everything below it optimizes for. The single hardest conversation in the book is also the one with the longest reach, which is why it repays the preparation it demands many times over, and why a leader who learns to have it well has acquired a skill that protects everything else they build.

The principle to carry out of the board conversation is that the second ledger is hard reality without a number, and your job upward is to supply the hardness the number would have. Translate every lever into risk, show the cost of the hedge against the cost of the failure, answer the three objections before they are raised, and give the board a question that makes them custodians rather than cutters. Win that, and the loop is safe from the one altitude that can kill it in a single meeting.

The board conversation is the last of the practical chapters, the how of putting the framework to work, from your first ninety days to your industry to your instruments to your failures to your boardroom. What remains is not another how. It is the why, and the who. Because all of this, the levers and the loop and the ledgers and the playbooks, finally asks something of you as a leader that no technique can supply, something that the whole book has been circling and has not yet named directly, and the closing chapter is about what that is.

> **THE EVIDENCE**
>
> Boards see only the first ledger by structure, not by failing: they govern through instruments, and the information that reaches them has already been compressed into metrics. The second ledger is invisible at that altitude. The task is not to make the board value soft things but to translate the second ledger into the board's native language of risk, cost, and competitive position.
>
> Frame every lever as a risk control and set the cost of the hedge against the cost of the failure it prevents. Answer the three standard objections: it slows the easy half to protect the hard half; the unmeasured risk is being carried regardless; and a better system has the same paradox, sharper. Convert the board into a custodian by reporting second-ledger leading indicators upward and planting the question, what is our system failing to see.

**MONDAY MORNING**

Before your next budget or board review, take your most vulnerable lever, the one most likely to be cut, and rewrite its justification entirely in the language of risk: name the specific expensive failure it prevents, estimate that failure's cost from your own history, and set it beside the small cost of the lever. Bring that one comparison, and the question, what is our system failing to see, rather than any appeal to treating people well.

PART FIVE

# Leading Through Paradox

CHAPTER SEVENTEEN

# Leading Through Paradox

*The paradox is permanent. The only variable is whether you lead it, or it leads you.*

Two years after the three phone calls, I got the four of us into one room, and for the first hour they barely talked about algorithms at all.

Marcus, Grace, and Priya had never met one another. I had connected each of them to the same idea without ever connecting them to each other, and now they sat across a table comparing notes, and what struck me most was how little they sounded like operators discussing systems and how much they sounded like leaders discussing people. They traded stories about specific human beings. They argued, warmly, about how much voice was too much. They finished one another's sentences across three industries that, on paper, had nothing whatsoever in common.

Marcus still talked about Carmen, two years on, the way a man talks about a debt he has paid but not forgotten. Grace talked about Ada and the near miss at bed nine. Priya talked about the outage at two in the morning and the engineer who could not find the bug in the system he had shipped. Each of them had started in precisely the same place, certain they had a technical problem with a technical fix, and each had arrived at precisely the same place, understanding that they had a human problem wearing a technical disguise, and that the disguise had nearly cost them the thing that mattered most in their organizations. The judgment, the voice, the growth of their people. They had each gone looking for a better machine and found, instead, a better way to lead.

This final chapter is about what they learned that went beyond the levers, the change in themselves that the levers both required and produced. Because you can install all five levers as a program and miss the entire point. The point is not the program. The point is a different way of leading, and it is available to anyone willing to hold a hard idea steady without flinching toward the easy one. The levers are how you do it. This chapter is about who you have to become to keep doing it when no one is making you.

At one point Grace asked the other two a question that stopped the room. "Would you go back?" she said. "If you could keep all the gains and undo all the work, rip out the forums and the coaching and the rotations and just run the systems the way you did at the start, would you?" None of them said yes. Not out of sentiment, but because they had each watched what the systems did when left to run alone, and they knew now that the gains and the rot came from the same place, and that the work was the only thing standing between them and a slow hollowing each of them had seen the early edge of. They had not added the levers to feel virtuous. They had added them because they had looked over the cliff.

## The One Thing They Shared

The deepest thing the three of them had in common by the end was a single shift, and it is the shift this whole book has been arguing toward.

They had stopped trying to fix the machine and started leading the people inside it. It sounds small. It is enormous, and it changes every decision that follows from it. A leader who is trying to fix the machine asks, how do I tune the system so it stops producing this harm. A leader who is leading through the paradox asks a completely different question. The system will keep producing this harm, because the harm is the price of the gain and the two cannot be separated, so what do I build around the system to keep my people whole. The first question has no good answer, which is exactly why the leaders who keep asking it stay stuck, tuning and retuning a dial that was never the problem. The second question has five answers, and they turn in a loop.

This is why two leaders can run the identical software, the identical dashboards, the identical tools, and end up with opposite organizations. The technology is the same. The posture toward it is not. One leader treats the system as the manager and spends their days adjusting it. The other treats the system as a powerful instrument that cannot see most of what matters, and spends their days building the human judgment around it that it cannot supply. The competence paradox is not, in the end, a problem of technology at all. It is a problem of leadership, and it is solved, to the degree it can be solved, by leaders who understand which half of the job is theirs.

## What I Got Wrong

I should be honest about my own part in this, because a book like this can sound as though the author arrived with the answer and handed it down, and that is not how any of it happened.

I did not design the five levers. I watched three leaders, in three industries, grope toward them under real pressure, and my contribution was mostly to notice that they were groping toward the same thing and to give the thing a name. The framework is a description of what worked, assembled after the fact, not a theory I imposed before it. That matters for how you should use it. These are not commandments. They are the distilled experience of people who were trying to keep good employees from quitting and good patients from being harmed and good code from rotting, and who discovered, separately and to their own surprise, that the same handful of moves kept working. Treat the levers the way they were found, as practical answers to be adapted, argued with, and improved by your own forum, rather than as a doctrine to be implemented and obeyed.

And I got things wrong along the way, which is worth saying because you will too. I underestimated how much the order of the levers mattered, and watched a leader fail by building a forum before there was any safety to fill it. I overestimated how much explanation alone would accomplish, the same error Priya made, until the evidence corrected me. The framework is more honest for having been wrong in public, and the propositions in the third chapter are written to be testable precisely so that you can find where it is wrong in your own organization, which it surely is in some way I cannot see from here. A framework that cannot be corrected by the people using it is just another black box, and we have spent a whole book on why those fail.

## The Five Disciplines

If I had to compress everything the three of them learned into a handful of disciplines a leader could carry into any algorithmic environment, into a warehouse or a ward or a codebase not yet built, it would be these five.

First, read both ledgers. The system keeps one, the measured one, and it keeps it beautifully, in high resolution, refreshed every few minutes. You must keep the other, the unmeasured ledger, the judgment and the voice and the learning and the trust that no dashboard records, because what you do not measure you will spend without ever noticing you are spending it. The discipline is to hold both ledgers in view at the same time and to trust the second one even though it has no numbers to defend itself with, especially because it has no numbers to defend itself with.

Second, diagnose the tension, do not chase the symptom. Turnover, a safety scare, a decaying codebase, a quiet floor, these are symptoms, and treating them directly always fails, because the cause is upstream. Underneath each symptom is one of the four tensions, specificity against autonomy, cadence against safety, efficiency against learning, transparency against trust, and the tension is the thing to address. The discipline is to ask, every single time a symptom appears, which of the four is producing it, and then to treat the structure rather than the surface, even though the surface is what everyone is pointing at.

Third, keep the gain and rebuild the space. The great temptation, every time, is to dial the machine back, and it never works, because it costs you the gain and does not restore the human capacity, which has withered from disuse and has to be rebuilt deliberately and slowly. The discipline is to refuse the false trade. Keep the precision, keep the speed, keep the efficiency, because they are real and valuable and someone else will keep them if you do not, and build, alongside them, the developmental space the machine compresses. Both hands working, always. Never the choice between the two that the machine keeps trying to make you make.

Fourth, run the loop, do not run a project. A project ends, and the ending is the trap, because the paradox does not end. The discipline is to convert your interventions from things you complete and celebrate into things you sustain and barely notice, a loop that turns on a steady cadence, forever, because a tension that never stops can only be met by a practice that never stops either. The leaders who fail here are often the ones who succeed first, who run a brilliant initiative, declare victory, and move on, and watch the paradox quietly return the moment the loop stops turning.

Fifth, protect the reserve. The judgment you do not need today is the judgment that saves you on the day the system is finally wrong, and that judgment exists on that day only if you spent the ordinary days maintaining it. The discipline is to invest, continuously and against the constant quarterly pressure, in the human reserve, the capability you hope never to need and cannot afford to be without, the pilot who can fly when the autopilot fails. It is the hardest discipline of the five, because it asks you to spend visible resources today against an invisible risk tomorrow, which is the one trade every measurement system on earth will tell you not to make.

Notice that the five disciplines are not independent, any more than the five levers were. They nest. Reading both ledgers is what lets you diagnose the tension rather than the symptom, because the symptom lives on the first ledger and the cause on the second. Diagnosing the tension is what lets you keep the gain and rebuild the space, because you cannot rebuild a space you have not correctly located. Keeping the gain and rebuilding the space is what the loop actually does, turn after turn. And protecting the reserve is the long discipline the loop exists to serve. They are five facings of a single way of seeing, and a leader who has truly absorbed one tends to find the others arriving on their own, because they were never really five separate things. They were one thing, looked at from five sides, which is fitting for a book about a paradox.

## The Leader's New Job

Underneath the five disciplines is a single redefinition of what a leader is actually for in the age of algorithmic management, and it is worth stating plainly, because most leaders are still operating on the old definition and quietly losing.

The old job of a manager was to do the things the system now does better. Assign the work, monitor the work, evaluate the work, coordinate the work. On every one of those, the machine wins, decisively, because it is faster and more consistent and tireless, and a leader who tries to compete there is competing with a calculator at arithmetic and will lose with dignity at best. The new job is to do the things the machine cannot do at all. To supply judgment where the system supplies only signal. To grant voice where the system grants only verdicts. To protect the developmental space the system erases. To keep the second ledger. To be, in short, the keeper of everything the system cannot see, which turns out to be most of what makes an organization survive contact with a bad day.

This is not a smaller job than the old one. It is a larger and harder one, and it is precisely the job the machine cannot take, because it consists entirely of the things the machine cannot do. The leaders who understand this are not threatened by the spread of algorithmic management. They are liberated by it, handed the dull and tiring half of their old job by a tireless assistant, and freed at last to do the human half that was always the actual point of leading. The ones who do not understand it spend their days trying to out-calculate the calculator, resenting the machine for being better at the part of the job that was never the important part, and they lose, slowly and then all at once, and the loss is invisible until the day the reserve they never built is the thing they suddenly need.

## A Word to the Skeptical Executive

Somewhere in this book a hardheaded executive has been waiting for the catch, suspecting that all this talk of voice and judgment and developmental space is the soft stuff that does not survive a bad quarter. Let me speak to that reader directly, because they are right to be suspicious and wrong about the conclusion.

Everything in this book pays for itself in the currency you already track, just on a delay that hides the payment. Carmen's departure cost Marcus a twenty-two percent retention collapse in her unit and a new hire gone in ninety days, all of it far more expensive than the calibration room that would have caught her. The near miss at bed nine was one alarm away from a harm event whose cost would have dwarfed a decade of coaching huddles. The outage Kofi could not fix threatened the entire revenue line of Priya's business, and the rotation program that eventually let him fix it cost a single week in six of one engineer's time. The levers are not charity, and they are not culture-building for its own sake. They are risk management for the specific, expensive, delayed failures that the systems quietly produce and hide, and they are cheaper than the failures by an order of magnitude.

The reason they look like soft costs is the very reason this whole book exists. The gains the systems produce are measured and immediate, so they are believed, and the costs are unmeasured and delayed, so they are doubted, and the skeptical executive is simply trusting the dashboard, which is precisely the trap. The hardest-headed thing a leader can do is to insist on paying a visible cost today to prevent an invisible one tomorrow, against the advice of every instrument in the building. That is not softness. It is the rarest form of discipline there is, and the leaders who have it will, in a decade, own the ground the coasting ones used to stand on.

## What This Asks of You

I will not pretend any of this is easy, because the leaders who found it easy were, without exception, the ones doing it wrong.

Holding a paradox is uncomfortable in a way that fixing a problem never is. A fix ends, and you get the clean satisfaction of completion, the box checked, the project closed. A paradox never ends, and so it offers no completion, only the low and permanent discomfort of holding two true things that pull in opposite directions and refuse to resolve. The human mind hates this. It wants the dial, the single answer, the project that closes and lets you rest. Leading through paradox means giving up the comfort of resolution and learning to live inside tension productively, which is a temperament at least as much as it is a technique, and it can be built, but it cannot be faked, and the people you lead will know the difference immediately.

And there will be pressure, constant pressure, to relapse. Every quarter the measured gains will be visible and the unmeasured costs will not, and someone reasonable and well-meaning will ask why you are spending real time and real money maintaining a reserve you are not currently using, running rotations that slow the shipping, holding forums that changed little this particular month. The discipline is to keep the loop turning through exactly that pressure, because the pressure is the paradox itself, speaking through a colleague, asking you politely to optimize the visible and abandon the invisible. That request is the original sin this entire book exists to help you recognize and refuse. It will never stop being made. Your job is to never quite stop refusing it.

## The Stakes Beyond the Organization

I want to end larger than I began, because the three phone calls that opened this book were never really about three companies.

Algorithmic management is not staying in the warehouse and the hospital and the software team. It is spreading into nearly every kind of work there is, and behind it the same kinds of systems are spreading into how we are governed, schooled, treated, judged, and informed. The competence paradox is not, in the long run, a management problem. It is becoming a human one. Everywhere a system optimizes a person, it builds one capability and risks quietly eroding another, and the question of whether we end up a society of people who can stand beside our machines as peers, or a society of people who have become their well-managed subjects, is being answered right now, one dashboard at a time, by leaders who mostly do not know that they are the ones answering it.

That is why the dedication of this book is what it is. The levers are small. The loop is modest. But multiplied across every organization that takes them up, they are a vote for a particular kind of future, one in which the astonishing power of these systems is bent toward making people more capable rather than less, more heard rather than more silent, more fully themselves rather than more interchangeable. That future is not guaranteed, and nothing about the technology delivers it automatically. It is a choice, and it is not made in legislatures or laboratories but in the ordinary, daily decisions of ordinary leaders about how the people in front of them will be managed. It is made in a thousand quiet rooms by people like Marcus and Grace and Priya. Which is to say, it is made by you.

## Start Small, Start Now

If all five levers and the loop sound like a great deal to take on, that is because, fully built, they are. So do not start there. Start with one.

You do not need the whole loop to begin. You need one lever, built honestly, on the tension that is hurting you most right now. If your best people are leaving, start with calibration, and give them a real hand in the targets. If your floor has gone quiet, start with hybrid coaching, and put a trusted peer beside the anxious. If your bench is thinning, start with rotation. Build that one lever well, close its loop, prove to your people that their voice changes something real, and you will have done the hardest part, which is never the mechanics but the credibility. Once people believe that engaging the system honestly leads somewhere, the second lever is far easier to add than the first was, because the trust the first one built is the soil the second one grows in.

And start now, because the costs compound the entire time you wait. Every quarter a system runs unaccompanied by these practices is a quarter of discretion eroded, candor lost, skill un-maintained, and trust spent, and none of it shows on a dashboard until it shows all at once. The leaders who began a year ago are not a year ahead. They are compounding-interest ahead, because the loop gets better the longer it turns, and a year of turning is a year their competitors cannot buy back at any price. The best time to have built the second ledger was the day you installed the system. The second best time is the morning you finish this book.

## The Last Word

Marcus, Grace, and Priya did not set out to answer a question about the future of human work. They set out to fix a turnover problem, a safety problem, and a quality problem.

They discovered, as you have, that underneath all three was the same paradox, and that meeting it well asked them to become better leaders than they had ever needed to be before. That is the gift hidden inside the competence paradox, and it is the note I want to end on. The machine, by taking over the easy half of leadership, has raised both the price and the value of the hard half. It has made judgment scarce, and therefore precious. It has made voice rare, and therefore powerful. It has made the leaders who can keep their people growing inside the optimizer into the most valuable leaders of the age, because they are doing the one thing that does not automate and never will.

I think often about the fact that none of the three of them set out to be the kind of leader this book describes. Marcus was an operator who loved a clean dashboard. Grace was a scientist who trusted her instruments. Priya was a builder who loved speed.

The paradox did not pick gentle people, or visionary ones. It picked competent ones, and it nearly broke each of them on exactly the strength they were proudest of, the operator's faith in the metric, the scientist's faith in the screen, the builder's faith in velocity. What they became was not softer. It was wiser, in the specific way that only a hard lesson about your own best instinct can make a person wiser. You do not have to be a different kind of leader to do this. You have to be willing to be corrected by the very thing you were most sure of, which is the hardest thing competence ever has to learn.

The paradox is permanent. That was never the bad news. It is the permanent invitation to lead, to really lead, in the one way the machine never can. The three calls came in a single quarter, and they were the same call, and it is the same call every leader will eventually have to answer. The only question the competence paradox finally asks is the one Marcus asked himself too late, standing in front of a red score that could not see the woman behind it. Not what does the system say. But what is it failing to see, and what will I do about it. Answer that question every day, with a turning loop and a steady hand, and you will lead your people well through the age of the machine. Fail to ask it, and the machine will answer it for you, the way it answered it for Carmen. The choice, as it has been since the very first dashboard glowed to life, is yours.

## THE EVIDENCE

Leading through paradox compresses into five disciplines: read both ledgers, diagnose the tension rather than the symptom, keep the gain while rebuilding the developmental space, run the loop rather than a project, and protect the human reserve. The five levers are how; the five disciplines are who you must become to keep using them under pressure.

The underlying redefinition is the leader's new job. The machine wins at assigning, monitoring, evaluating, and coordinating, so the leader's work becomes everything the machine cannot do: judgment, voice, developmental space, and the keeping of the second ledger. It is a larger job than the old one, and the only one that does not automate.

## MONDAY MORNING

Tomorrow, before you look at a single dashboard, walk to where your people actually do the work and ask one of them what the system cannot see about their job. Write down the answer, and put it somewhere your metrics cannot reach but your decisions can. Do that once a week for a year. You will have built, without any technology at all, the second ledger, and the habit of reading it is very nearly the whole of leading through the paradox.

# Appendices and Reference

APPENDIX A

# The Four Propositions

The argument of this book rests on four falsifiable propositions, drawn from the underlying review. They are stated here formally, each with a way to test it in your own organization, because a claim you cannot test is only a belief, and this book is trying to give you something sturdier than a belief.

## Proposition One: Specificity and Autonomy

Where an algorithmic system issues highly specific guidance that refreshes rapidly, on the order of every few minutes, and the worker's technology readiness is low, perceived autonomy falls by roughly four tenths of a standard deviation even as task self-efficacy rises by about three tenths. Higher technology readiness, at or above the fourth point of a five-point scale, buffers the autonomy loss. Accuracy rises by about twelve percent while independent exploration falls by about thirty percent, and the felt sense of freedom drops sharply after roughly five consecutive prompts.

How to test it. Measure three things: the refresh cadence of your system's guidance, a short technology-readiness score for each worker, and a brief autonomy scale. Compare high-readiness and low-readiness workers under the same cadence. If the proposition holds, the autonomy gap between them will widen as the cadence quickens, and you will see accuracy and exploration moving in opposite directions at once.

## Proposition Two: Cadence and Safety

Where the system issues more than thirty prompts per hour and transparency is low, psychological safety falls by roughly half a standard deviation, and the rate at which workers voice concerns or dissent falls by about a quarter.

How to test it. Count prompts per worker per hour. Measure the rate of voiced concerns, near-miss reports, or open dissent in team settings. Compare a high-cadence, opaque condition against a lower-cadence or explained one. A drop in voiced concern under high opaque cadence, even as output holds steady, is the signature of this proposition at work.

## Proposition Three: Efficiency and Learning

Where the system delivers twenty or more optimization nudges per shift and the worker's learning orientation is low, exploratory behavior falls by about fifteen percent while measured productivity rises by roughly eighteen percent. Workers high in learning orientation resist the exploration loss.

How to test it. Count optimization nudges per shift. Use a short learning-orientation measure. Track an exploration proxy, such as attempts at novel methods or the use of non-prescribed approaches. Segment by learning orientation. If the proposition holds, low-orientation workers will show the exploration drop and high-orientation workers will not, at the very same nudge rate.

## Proposition Four: Transparency and Trust

Where the system's logic is disclosed, workers have writable panes through which they can annotate or contest its outputs, and the setting is relatively low in power distance, trust in the system rises by about half a standard deviation. Disclosure without the ability to respond leaves trust unchanged. The presence of annotation rights raises acceptance of the system's outputs by roughly forty percent.

How to test it. Run a simple comparison. Give one group explanation alone, and another explanation plus a writable pane whose entries are visibly acted upon. Measure trust and acceptance in both. If the proposition holds, explanation alone will move comprehension but not trust, and the group with real voice will show the trust and acceptance gains, most strongly where people feel free to contradict authority.

Treat these as predictions, not as doctrine. If your organization contradicts one of them, you have learned something the review could not have told you about your own conditions, and you should trust your data over this book. The propositions are written to be falsifiable precisely so that you can find out where they fail for you.

APPENDIX B

# Implementation Checklists

These checklists compress Part Three into something you can hold in one hand during a planning meeting. They are starting points, not scripts. Adapt them in your own words once your voice forum is running.

## Lever One: Goal Calibration

- Identify which metrics carry real choices and are therefore contestable, and name plainly the few that are fixed by law, safety, or strategy.
- Bring the people the metric measures into the room, mixing veterans and newcomers, and present the metric as a draft to improve rather than a verdict to accept.
- Ask the three questions: what helps, what insults your judgment, and what is the system missing.
- Change what you can on the spot, and explain specifically what you cannot, and why, in the room and on the same day.
- Run it on a standing cadence, and open every session by reporting what changed because of the last one.

- Wall the calibration room off from evaluation, so that what people say about the metric can never affect their own rating.

## Lever Two: Hybrid Coaching

- Pair an experienced peer, never the worker's evaluator, with the worker for regular, protected coaching time.
- Coach by asking, not telling: what is the system seeing, what might it be missing, what do you think.
- Use all three modes: in the moment for high-stakes signals, in review for skill-building, in pattern to catch drift toward over-compliance or over-dismissal.
- Let the algorithm triage where a human conversation will pay off most, so scarce coaching attention goes to the moments that matter.
- Feed the misreads coaches keep finding back into goal calibration as system-level signal.
- Treat coaching as a skill people rotate into, so the practice replenishes its own coaches.

## Lever Three: Rotational Upskilling

- Identify the few load-bearing skills your operation cannot survive without on its worst day.
- Schedule protected, recurring practice in those skills with the tools turned down, and protect it like a customer commitment.
- Frame it as a simulator, training rather than punishment, so that needing practice carries no stigma.
- Use real, hard, consequential problems, because practice with nothing at stake builds nothing.
- Make it count as visible, valued work, or it will be the first thing cut under deadline pressure.
- Do not over-rotate; you are buying insurance, not refusing the tool.

## Lever Four: Explainable Dashboards, the CLEAR Test

- Context: state why this signal is appearing, and why now.
- Logic: show the reasoning chain a human can follow, not the raw weights.
- Evidence: show the specific data, so the worker can check it against reality.

- Action: tell the worker what they can do, including how to disagree, and route the disagreement somewhere real.
- Recap: give a short, plain, actionable next step.
- Test it by asking a worker to explain the flag back to you; if they cannot, it is not CLEAR yet.

## Lever Five: Voice Forums

- Stand up a moderated, recurring forum with real authority to change something about the system that manages people.
- Rotate membership and cross levels deliberately, and work hardest to include the people who have gone quiet.
- Bound the scope explicitly: name what the forum can change and what it cannot.
- Have the moderator synthesize many voices into coherent proposals and protect the quiet from the loud.
- Close the loop every time, reporting what changed because of the last session.
- Start with a scope so small it is almost embarrassing; the reliability of the loop closing matters more than its size.

## Running the Loop

- Wire the levers together: calibration sets the targets, dashboards deliver them clearly, coaching interprets them, rotation keeps the skill to act, and forums feed everything back to calibration.
- Diagnose by tracing symptoms upstream: a forum that has gone quiet often means a weaker lever somewhere else in the loop.
- Keep it turning through quarterly pressure; the danger is not sudden collapse but a slow stall that no one notices.
- Start with one lever on your worst tension, prove that voice is real, then add the next.

APPENDIX C

# Finding Your Tensions

Use this to locate which of the four tensions is doing the most damage in your organization right now, so that you know which lever to build first. Answer honestly, and answer from evidence where you have it rather than from impression.

## Signs of Specificity Against Autonomy

- Your best and most experienced people are the ones quietly disengaging or leaving.
- Workers follow the system's guidance even when they can plainly see it is wrong.
- People describe their work as following instructions rather than exercising judgment.
- Performance scores are up while ownership and initiative are visibly down.

If several of these are true, start with goal calibration.

## Signs of Cadence Against Safety

- The floor has gone quiet, and people raise fewer concerns than they used to.
- Workers either chase every alert or have learned to dismiss alerts wholesale.

- Near misses are caught late, or not reported at all.
- The faster the system's feedback runs, the less anyone says out loud.

If several of these are true, start with hybrid coaching.

## Signs of Efficiency Against Learning

- Junior people are productive immediately but cannot handle the non-routine case.
- When the system fails, fewer and fewer people can step in to fix it.
- Your senior bench is thinning and is not being replenished from within.
- Output is high while the underlying skill is visibly fading.

If several of these are true, start with rotational upskilling.

## Signs of Transparency Against Trust

- People understand a system perfectly well and still do not trust it.
- Workers game the metric, route around the system, or feed it what it wants to hear.
- Explaining the system more has not improved trust, and may have lowered it.

- There is no real way for a worker to contest the system and be heard.

If several of these are true, start with explainable dashboards and voice forums.

Most organizations have more than one of these at once, because the four tensions share their roots. Begin with the one whose signs are sharpest, build its lever well, and let the loop pull the others into reach.

# Glossary

The key terms of the framework, gathered in one place for reference.

**Competence Paradox.** The central idea of this book: that algorithmic systems build one kind of competence while eroding another, through the same mechanism and at the same time, with the erosion hidden on the very dashboards that produce it.

**Competence Surge.** The genuine, measured rise in accuracy, speed, or consistency that a system produces, and the half of the paradox that every dashboard can see.

**Discretion Erosion.** The quiet loss of a worker's judgment and ownership as a system narrows the decisions left to them, the unmeasured half of the specificity tension.

**Mastery Signal.** The reassurance a fast, confident system gives that the work is under control, which can substitute for the worker's own developing sense of mastery.

**Spotlight Threat.** The felt exposure of working under rapid, visible monitoring, which suppresses the voicing of concern and the admission of uncertainty.

**Success Cue.** The immediate positive feedback an optimization system supplies, which can crowd out the slower, harder practice through which deep skill is built.

**Practice Crowding.** The displacement of effortful learning by efficient assistance, the mechanism behind the efficiency-against-learning tension.

**Explanatory-Assurance.** The relief of no longer facing a mystery that disclosure provides, a real good that answers the question do I understand, but not the question of standing beneath it.

**Participatory-Validation.** The confirmation that one's judgment counts, which only voice can supply, and the thing that actually builds trust in a system.

**The Second Ledger.** The unmeasured account of judgment, voice, learning, and trust that no dashboard records, and that a leader must keep deliberately because what is not measured is spent unnoticed.

**CLEAR.** The structure of an explainable dashboard: Context, Logic, Evidence, Action, and Recap. The Action step, which supplies recourse and voice, is the one most systems omit and the one that builds trust.

**Dynamic Equilibrium.** Paradox theory's answer to a tension that cannot be solved: a balance actively and continuously maintained, like a cyclist staying upright only by steering. The five-lever loop is this idea made operational.

**Goal Calibration.** The first lever: bringing the people who do the work into the setting and adjustment of the targets the system optimizes, with the leader still holding the pen.

**Hybrid Coaching.** The second lever: pairing algorithmic feedback with human coaching that helps a worker interpret the signal and decide, reinstalling the judgment step the system removes.

**Rotational Upskilling.** The third lever: scheduled, protected practice in the load-bearing skills the system has automated, so they do not vanish from disuse.

**Voice Forum.** The fifth lever: a moderated, standing channel through which the people under a system shape its ongoing design, distinct from calibration in that its scope is the whole system.

**Technology Readiness.** A worker's disposition and capability to engage with new technology, a moderator that buffers the autonomy loss of precise, fast systems.

**Power Distance.** The degree to which people in a group accept and expect steep hierarchy. Low power distance lets writable panes and voice forums work; high power distance leaves them unused until leaders prove dissent is safe.

**Psychological Safety.** The shared belief that it is safe to speak up, take risks, and admit error, eroded by high-cadence opaque monitoring and rebuilt by coaching that is walled off from evaluation.

**Learning Orientation.** A worker's disposition to seek mastery and treat difficulty as growth, which protects exploratory behavior against the crowding effect of constant optimization.

# Acknowledgments

This book exists because of people who gave it more than I had any right to ask.

My deepest thanks go to the researchers whose field studies, gathered patiently across many years and many workplaces, are the true foundation of everything here. The frameworks in this book only summarize what they discovered in the difficult, unglamorous work of observing real people under real systems, and the lessons were theirs long before they were ever mine. A synthesis is only as good as the evidence it rests on, and I was fortunate in mine.

To the clients and colleagues across retail, financial technology, artificial intelligence, and manufacturing who let me sit in the rooms where these decisions were actually made, and whose experiences, disguised throughout these pages to protect their confidence, taught me what no dataset could. You know who you are, even if no one else does, and I am grateful beyond what these few lines can hold.

To the mentors and partners who saw the argument before it was an argument, and who pushed me to make it sharper, clearer, and more honest than I would have managed alone. To the early readers whose criticism saved the book from the mistakes that would have embarrassed it most, and who were generous enough to tell me the hard things while there was still time to fix them. A book is improved most by the people willing to say what is not yet working, and I was fortunate in mine.

To my colleagues at Ideashaus Consulting, for the conversations, the patience, and the time that made the writing possible alongside the work, and for tolerating a founder who kept disappearing to wrestle with a manuscript when there were clients to serve. The firm's work is the soil this book grew in, and every idea here was tested first against the unforgiving reality of real organizations trying to lead people well through a changing world.

And to my family, who endured the long, quiet, unrewarded middle of writing a book about the things that go unmeasured, and who never stopped believing the road led somewhere. It did, and you are the proof.

Any insight in these pages belongs partly to all of you. The errors are mine alone.

# A Note on Sources

A book about how systems measure people owes its readers an honest account of how its own evidence was assembled, because the method matters as much as the conclusions, and because a reader has every right to know which of these stories are drawn from research and which are something else.

The material in this book comes in three kinds, and I have tried throughout to make clear which kind you are reading.

The first is the evidence base. The spine of the argument is a systematic scoping review of eighty-two peer-reviewed field studies of algorithmic and AI-assisted management, spanning logistics, software, healthcare, finance, and retail, synthesized through the lens of paradox theory. That review is my own peer-reviewed work, published in Human Resource Development Review in 2025. The prevalence figures, the four tensions, the four mechanism pairs, and the four propositions, including their effect sizes and thresholds, all derive from it and are reported here as they were found there. A reader who wants the full corpus, the inclusion criteria, and the study-by-study evidence should begin with that review, to which the Notes that follow point chapter by chapter.

Four theoretical traditions run underneath the whole framework, and a reader new to them will find the book richer for a little time spent with the originals. Goal-setting theory, which explains why specific targets drive performance and at what cost, comes from the long program of work by Edwin Locke and Gary Latham. Self-determination theory, which gives us the needs for autonomy, competence, and relatedness, comes from Edward Deci and Richard Ryan. The job demands-resources model, which explains how the pace and design of work produce strain or engagement, comes from Arnold Bakker and Evangelia Demerouti. And paradox theory, which gives us the central idea that competing demands are managed rather than solved, through a dynamic equilibrium, comes from the work of Wendy Smith, Marianne Lewis, and their colleagues.

The second kind is the composite. The leaders and workers who carry the book, Marcus and Grace and Priya, and the people around them, are composites, built from the situations documented in the eighty-two studies and from my own work advising companies across retail, financial technology, artificial intelligence, and manufacturing. The underlying dynamics are real, but the names, the identifying details, and sometimes the specific sequence of events have been changed, combined, or obscured to protect the confidentiality of the people and organizations involved. A reader should not try to identify the real companies behind them, because the disguise is deliberate and the identification would in most cases be wrong. What I have preserved in every composite is the truth of the lesson, the actual dynamic that the research documents, because that is the part worth keeping. The disguise protects the people; the lesson is faithful to what really happens.

The third kind is the openly illustrative scenario, invented to make a principle visible rather than to report an event, the way a physics problem invents a frictionless plane. Where I recount a brief case to demonstrate a single dynamic, I am isolating a pattern from the noise that surrounds it in any real workplace. I have tried to flag these clearly, so that no reader mistakes an illustration for a finding. They are true in the way a parable is true, faithful to how the pattern works without claiming that any particular instance occurred.

I have mixed these three kinds because each does something the others cannot. The research lets the reader verify the claims. The composites let me show the dynamics as they actually play out in real workplaces without betraying a confidence. The illustrations let me isolate a principle from the surrounding noise. Used together, and labeled honestly, they let the book be both vivid and trustworthy, which is the only combination worth aiming for in a book that asks you to act on what it claims. The honest reader's rule applies throughout: trust the evidence over the illustration, and trust your own organization's data over both.

# Notes

The notes that follow indicate, chapter by chapter, the evidence on which each rests. They point to the underlying review for the empirical claims, name the theoretical traditions that inform each argument, and flag which figures are composites. They are a guide for the reader who wishes to verify or read further, not exhaustive academic citations; the full scholarly apparatus is in the underlying review. Figures cited in the text are given as reported there.

Prologue and Chapter One. The three leaders, Marcus, Grace, and Priya, and the opening scenes that introduce them, are composites, as described in the Note on Sources. The account of how algorithmic management has spread across industries draws on the corpus of eighty-two field studies assembled in the review. Marcus and Carmen are composites; the warehouse retention figure illustrates the pattern the review documents rather than reporting a single firm. The framing of the dashboard boss is the author's.

Chapter Two. The central paradox, the image of the two ledgers, and the mechanism framing are the author's synthesis of the review, grounded in paradox theory (Smith and Lewis), which holds that competing demands are managed in a dynamic equilibrium rather than solved.

Chapter Three. The four tensions and their prevalence across the corpus, specificity and autonomy at roughly fifty-eight percent of studies, cadence and safety at forty-seven, efficiency and learning at forty-four, and transparency and trust at forty-two, are reported as found in the review, along with its scoping-review method.

Chapter Four. Proposition One and its effect sizes, the autonomy loss of about four tenths of a standard deviation, the self-efficacy gain of about three tenths, the buffering role of technology readiness, and the trade between accuracy and exploration, are stated as reported in the review. The argument is grounded in goal-setting theory (Locke and Latham) and self-determination theory (Deci and Ryan). Marcus, Daniel, and Rosa are composites.

Chapter Five. Proposition Two, the prompt-cadence threshold and the effects on psychological safety and voiced dissent, is as reported in the review. The argument is grounded in the job demands-resources model (Bakker and Demerouti) and the literature on psychological safety. Grace, Ada, Bisi, and the bed-nine episode are composites.

Chapter Six. Proposition Three, the optimization-nudge threshold and its effects on exploration and productivity, together with the moderating role of learning orientation, is as reported in the review. The argument draws on self-determination theory and the literature on workplace learning. Priya, Kofi, and Lena are composites.

Chapter Seven. Proposition Four, the roles of disclosed logic, writable panes, and power distance, the inertness of explanation offered alone, and the effect of annotation rights on acceptance, is as reported in the review. The composites are as above.

Chapters Eight and Nine. The first two levers, goal calibration and hybrid coaching, synthesize the review's findings on participative goal-setting and human-in-the-loop feedback; goal-setting and self-determination theory inform them. The recounted cases of Jo and of Marisol are illustrative scenarios, as described in the Note on Sources.

Chapters Ten and Eleven. The third, fourth, and fifth levers, rotational upskilling, explainable dashboards, and voice forums, and the loop that joins all five, are the author's synthesis of the review's findings. The CLEAR structure for explainable dashboards is the author's framing of those findings.

Chapters Twelve through Sixteen. These implementation chapters, on the first ninety days, on industry playbooks, on measuring the second ledger, on the ways the work fails, and on the board conversation, extend the framework into practice. They draw on the same review and on the author's consulting work across retail, financial technology, artificial intelligence, and manufacturing. All named figures in them are composites, and the brief cases are illustrative. The application of Goodhart's law to the second ledger, in the chapter on measurement, follows the familiar formulation that a measure which becomes a target ceases to be a good measure.

Chapter Seventeen. The closing synthesis, and the convergence of the three leaders on a single practice, is the author's, drawn together from the whole of the preceding argument.

# About the Author

Ofem Ekapong Ofem is the founder of Ideashaus Consulting, where he advises companies in retail, financial technology, artificial intelligence, and manufacturing on leadership, technology, and the human side of organizations being reshaped by intelligent systems. Over more than a decade of work across those industries, he has sat in the rooms where the decisions in this book were made, watching people grow more capable and less heard under the systems built to manage them, and the lessons of those engagements, disguised to protect confidence, run through every chapter.

His research on the competence paradox, the central idea of this book, appears in Human Resource Development Review. He works with founders, executives, and boards on the defining management problem of the age of artificial intelligence: how to lead people well when a system, and not a manager, increasingly sets the pace and shape of their work.

He is the author of several previous books, among them The Advantage Window: Why the Best Technology Loses, and What Wins Instead in the Age of AI; Unbreakable: Why Resilience Is a Design Problem, Not a Test of Strength; and Rewire: Thirty-One Proven Hacks for Co-Existing with AI. He lives in Minneapolis.

Made in the USA
Monee, IL
05 November 2026

67811014R00190